BASIC ENGLISH FOR BUSINESS AND TECHNICAL CAREERS

BASIC ENGLISH FOR BUSINESS AND TECHNICAL CAREERS

Leila R. Smith
Los Angeles Harbor College

John Wiley & Sons
New York Chichester Brisbane Toronto Singapore

Library of Congress Cataloging in Publication Data:

Smith, Leila R., 1928–
Basic English for business and technical careers.

An abridgment of: English for careers.
Includes index.
1. English language—Business English. 2. English language—Technical English. 3. English language—Grammar—1950– . I. Smith, Leila R., 1928– English for careers. II. Title.
PE1115.S6213 1985 428.2'02465 84–25617
ISBN 0–471–81077–0

Printed in the United States of America

10 9 8 7 6 5 4 3

To Seymour, Eric, Roberta, Udaya, Sean, and Sarala Rose

A word fitly spoken
is like apples of gold in settings of silver.
—the Bible

TO THE INSTRUCTOR AND TRAINING DIRECTOR

Basic English for Business and Technical Careers is an abridgment of *English for Careers: Business, Professional, and Technical.* The instructor's edition is the same for both books and may be obtained by writing on your letterhead to—

John Wiley & Sons, Inc.
605 Third Avenue
New York, NY 10158

For adapting this text to individualized instruction, see page 75 of the "Instructor's Supplement."

The comments and suggestions made by many of you contributed to this realistic and practical approach to developing the language skills of our students. I am grateful for your help. Among the many educators I wish to thank are Ellen D. Alderman, Moultrie Technical Vocational School; Joelene L. Mack, Golden West College; Edna V. Jellesed, Lane Community College; Carol Ann Pitz, Kensington Business Institute; Joanne Murcar, Spokane Falls Community College; Dolores Denova, Los Angeles Harbor College; Betty M. Dooley, Clark Technical College; Joy D. Schuhmann, Internation Business College; Jo Ann Sumner, Sierra College; Joan Teufel, Rio Hondo Community College; Nita Wade, Louisiana Business College; Sheila L. Long, Miami-Dade Community College; Quilvie G. Mills, Miami-Dade Community College; Dorothy Prevatt, Miami-Dade Community College; Ruth Sanford, Spokane Community College; Loretta J. Drummond, Strayer Business College; Joan Sterns, Southeast Community College; H. Grace Heringer, Skyline College; Norma R. Vara, Laredo Jr. College; John Holmstedt, Clatsop Community College; Margaret Helen Taylor, Fountain Valley, CA; Lucie Givins, Heald College.

If you have questions or comments about the *Basic English for Business and Technical Careers* course, write or call me at Los Angeles Harbor College, 1111 Figueroa Place, Wilmington, Caifornia 90744.

Leila R. Smith

ABOUT THIS BOOK

This book is different. . . . You don't browse through it. You don't read it like other books. What you DO is *learn* your way through it!

Each of the 10 sections has a starting page or two followed by READ AND REPLAY exercises. In these unique *learning units*, the information to be learned is arranged step by step in short portions. These learning units present the information and then give you an opportunity to "replay" what you learned. The "replay" enables you to apply the newly learned skill immediately. As soon as you complete a REPLAY, you check the answers in the back of the book. Students have been extremely enthusiastic about this new way to learn.

Not only do you learn and retain more, you also enjoy the process. Because doing the REPLAYS is interesting and challenging, however, some students are tempted to pretest their English knowledge by doing the REPLAY exercises without reading the material in the READ unit. You should resist doing this for the following reasons:

1. The formula—first, read; second, respond; and third, see if you match—has been proven effective through research and testing. If you violate this pattern, you learn substantially less and are not as well prepared to take the quizzes.
2. You miss the personal satisfaction that comes from making all (or nearly all) correct responses.
3. You make more mistakes and, as a result, you lose rather than save time.

Why is READ AND REPLAY a good learning system?

- By quickly making active responses, you immediately apply what you learned; therefore, you understand it better and retain it longer.
- The small learning units are more efficient than longer ones.
- The READ AND REPLAY units hold your attention better, and your powers of concentration improve.
- READ AND REPLAY involves you to the point that you are motivated to continue beyond normal attention spans.

Other special features of READ AND REPLAY are the learning goals provided at the beginning of each section. Read these starting pages carefully, as they enable you to find out exactly what skills and knowledge you will master by the time you complete each section. The READ AND REPLAY units come next. Finally, Checkpoint, Special Assignment, Proofreading for Careers and Practice Quiz round out each section.

While studying *Basic English for Business and Technical Careers*, you also learn more about the world of business. This side-by-side learning happens because many of the sentences selected to illustrate English points deal with business matters.

Enjoy READ AND REPLAY. Give it a chance, and you'll find that your command of English will be a lifelong asset to your career.

Leila R. Smith

Contents

SECTION 7 Love Lines 165

(Adjectives and Adverbs)

SECTION 8 The Taming of the Apostrophe 189

SECTION 9 The Pause That Refreshes 205

(Commas)

SECTION 10 Punctuation Potpourri 227

(. ? ! ; : " - ' --)

BASIC ENGLISH FOR BUSINESS AND TECHNICAL CAREERS

PRETEST

Read the following sentence before answering No.'s 1 and 2.

FEW ECONOMISTS BELIEVE THAT PRIVATE INDUSTRY WILL DEVELOP METHODS FOR REDUCING INFLATIONARY PRESSURES.

_____ 1. In the sentence shown above, INFLATIONARY is used as a/an (a) verb (b) adjective (c) noun (d) adverb (e) preposition.

_____ 2. In the sentence shown above, FOR is used as a/an (a) verb (b) adjective (c) conjunction (d) preposition (e) adverb.

_____ 3. If the first line of the inside address of a business letter is "Mr. Samuel E. Smith," the preferred salutation would be (a) Gentlemen: (b) Dear Sir: (c) Dear Sirs: (d) Dear Mr. Smith: (e) Dear Mr. Samuel E. Smith:.

_____ 4. PER DIEM means (a) by the person (b) by the day (c) for the year (d) for God and country (e) punched card is needed.

_____ 5. These (a) phenomena (b) phenomenae (c) phenomenon (d) phenomeni (e) phenomenas have never before occurred.

_____ 6. The (a) General Manager (b) general manager (c) General manager (d) general Manager of the

_____ 7. new (a) Company (b) company is a wealthy woman.

_____ 8. After three (a) year's (b) years' (c) years work in

_____ 9. the (a) children's (b) childrens' (c) childrens department, she resigned.

_____ 10. Please give the reports to Ozelle and (a) I (b) me (c) myself.

_____ 11. Ms. Denova is the one (a) who (b) whom we believe will be offered the job.

_____ 12. If you had (a) drive (b) drove (c) driven to work this morning, you would have seen the accident.

_____ 13. Which would be (a) easyer (b) easiest (c) easier (d) easyest for you to prepare, a letter or a short report?

_____ 14. We hope to receive (a) a (b) an 18 percent discount.

_____ 15. AFTER WORKING IN THE STORE ALL DAY, HIS FEET WERE TIRED. The preceding sentence is an example of a (a) fragment (b) sentence that lacks parallel construction (c) sentence with a misplaced part (d) sentence with a dangler (e) correctly written sentence.

For No.'s 16–20, answer (a) if all the punctuation is correct. If there is any kind of punctuation error, answer (b).

_____ 16. "The new office furniture," he said, "is too expensive".

_____ 17. Money, beauty, intelligence, and charm—she has all of them.

_____ 18. Mr. Crane is not here, however, I can help you.

_____ 19. We've mailed you a copy of the new book and hope it will reach you before the end of the month.

_____ 20. Will you please send us your check for our Invoice No. 4268 within ten days.

After you find out how many you answered correctly, see the reverse side of this page for your fortune.

No. Right

18–20	You have a good command of English for your career. This course will serve as a brushup, and you'll easily achieve expertise.
15–17	Your business English skill is fair. After the practice provided in this course along with the review of some rules you've forgotten, you'll be an expert in the English required for your career.
12–14	Your English for business needs improvement. Because your basic language skill is good, you'll enjoy studying and mastering the skills that will make you an expert.
0–11	You came to the right place! Taking this course is a wise decision. You'll find the instruction extremely informative. Because you are now motivated to learn English for your career, you'll capture those principles that escaped you in the past. YOUR EFFORT IN THIS COURSE = SUCCESS.

THE FRESH START

After Completing This Introduction, You Will:

- *Be confident about the success you will achieve in this English course.*
- *Know what is meant by "English for careers."*
- *Understand why this book is the realistic and businesslike way to learn English for careers.*

In the year 1414, Sigismund, Emperor of the Holy Roman Empire, said to a prelate who had objected to His Majesty's grammar: "Ego sum rex Romanus et supra grammaticam." (I am the Roman king and am above grammar.)

If *you* happen to be a Roman king, you don't need to read on.

For the rest of us, the type of language we use, both spoken and written, significantly affects how successful we are at earning a living and getting along with other people. For example, language that might contribute to the success of a construction foreman giving orders to his crew might be quite unsuccessful when used by a bank manager speaking to a customer—and vice versa.

THE LANGUAGE OF CAREERS

You have probably been using English all your life to help communicate your ideas to others. You will, of course, not learn a new language in this course, because "career English" is not a special or separate language. The work in this book provides a review and reinforcement of the principles of English usage that you have studied throughout your years of formal education. When you have completed the course, following the instructions given in this book and by your instructor, you should be confident of your ability to . . .

Speak and write the style of English that is most likely to lead to success in a business, professional, or technical career.

This language style is known as **Standard English**, which is . . .

The language that is written and spoken by well-educated Americans in both formal and informal situations.

The style of English that you will study is not based on a long list of rules taken from traditional grammar books. Instead, you will be asked to learn only those principles that reflect the actual speaking and writing styles of well-educated and well-informed people today. The emphasis is on those aspects of Standard English that commonly appear in business and professional communications and that are often misused by students. Most people need instruction in Standard English for careers to be sure of being "right."

OTHER TYPES OF LANGUAGE

Some of you might use a nonstandard type of English in your everyday conversation. You do not need to eliminate this other type of English; you might choose to use it in appropriate situations. Switching types of usage or language is a common occurrence—students or businesspeople may use slang with certain friends or a neighborhood dialect with friends and relatives, but they may then switch to Standard English with other friends or business and professional acquaintances.

THE ENGLISH FOR CAREERS SYSTEM

English for Careers is the realistic and businesslike system of learning standard English for your career because:

1. Many grammar terms have been simplified and others have been left out.
2. Many traditional grammar rules have been omitted, since a number of well-educated Americans no longer apply them in their speech and writing.
3. READ AND REPLAY units of instruction along with frequent quizzes make learning efficient. Your success on the quiz at the end of each section provides constant assurance that you are achieving your goals.

WHAT WILL YOU LEARN?

You learn only the grammar terms and principles that meet at least two of these three standards:

1. Understanding the term or principle enables you to correct a commonly made error.
2. Learning the correct form can be accomplished without understanding complicated grammar terminology.
3. Using the incorrect form is noticeable to many people who use Standard English.

CHALLENGE FOR THE EXPERT

Some of you have previously mastered grammar terminology as well as many of the traditional grammar rules and now wish to strive for PERFECTION in usage. Therefore, a Grammar Supplement is provided to enable you to review traditional grammar instruction that is omitted from the main part of the book. This supplement will also serve as a reference when you want to be absolutely certain of a particular usage.

Remember—the Grammar Supplement will *not* make the work in the

READ AND REPLAY sections easier to complete. Instead, the supplement provides additional challenge or enrichment of instruction for those who might profit from it. Follow your instructor's suggestions regarding its use.

REPLAY

Determine your understanding of what you have read by answering these questions. If you are unsure of any of your answers, reread pages 3 and 4.

1. The language style most appropriate for business and professional careers is known as ______________________.
2. It's wrong to use slang when writing a letter to a friend. T___ F___
3. You should be sure to use the same style of language in all your communications. T___ F___
4. Standard English is always superior to other types of English. T___ F___
5. Standard English is controlled by a long list of grammar rules that well-educated people strictly obey. T___ F___
6. If you want to pass this course, be sure to refer frequently to the Grammar Supplement. T___ F___
7. By carefully completing the 10 sections of this book, you will achieve thorough mastery of traditional grammar terminology and rules. T___ F___
8. By carefully completing the 10 sections of this book, you will achieve confidence in your ability to speak and write English in the style most likely to lead to success in a business, professional, or technical career. T___ F___

Please see page 307 to check your answers.

SECTION 1

TOOLS OF THE TRADE

After Completing Section One, You Will:

- *Identify a part of speech by how the word is used in a particular sentence.*
- *Demonstrate sufficient familiarity with the parts of speech so that you can review the Standard English principles presented in this book.*

Capable carpenters identify and use correctly the tools necessary for that trade. It is not, however, necessary for them to know everything about the construction and manufacture of those tools.

If you want to teach someone to type by touch, you first show your student a few basic parts of the typewriter—the carriage return, the paper release, the space bar, and so on. For example, the learner must know where the space bar is before you tell him to "depress the space bar with the right thumb." Although some typists might be interested in understanding HOW a typewriter works, this knowledge is not necessary for becoming an expert typist.

Should you decide to become a linguistics expert or an English teacher, you would pursue an in-depth study of English grammar. For mastering English for most careers, however, you just need to know how to identify and use correctly the basic tools of the English "trade." These tools are called the "parts of speech." The ability to identify some of the parts of speech in certain sentences will enable you to understand the explanations that follow in later chapters. That is, a minimum grammar vocabulary is presented in this section so that, for example, when the word "pronoun" is used by your instructor, by a student, or in this book, the word will mean the same for all.

The parts of speech reveal the system for placing words into various categories. Every word in our language belongs in one of these categories. Most of you have studied the parts of speech in the past. This section is intended as a review. You'll find additional details on parts of speech in Sections Four, Five, Six, and Seven and in the Grammar Supplement.

READ UNIT 1

NOUNS AND VERBS

NOUNS name persons, animals, things, ideas, places, times, activities. Here are examples of various kinds of nouns and how they are used in sentences.

1. **Persons**—man, John, accountant, members, lady, father.

   My *father* spoke to the *members.*
   The *man* named *John* is studying to be an *accountant.*

2. **Animals** (fish, birds, insects, etc.)—goldfish, dog, Fido, cockroach, eagle, dinosaur.

   Fido is afraid of *goldfish.*
   Dinosaurs were prehistoric *animals.*

3. **Things**—desk, pencil, merchandise, machines, picture, telephone.

   We have two *machines* that process the *merchandise.*
   He put the *pencil,* the *telephone,* and the *picture* on the *desk.*

4. **Ideas**—honesty, beauty, democracy, evil, kindness, modesty.

   My *beauty* is exceeded only by my *modesty.*
   His *honesty* cannot be questioned, but he lacks *kindness.*

5. **Places**—home, Illinois, moon, kitchen, city, factory.

   My *home* in *Illinois* has a modern *kitchen.*
   We plan to build a *factory* in a *city* on the *moon.*

6. **Times**—birthday, Saturday, Christmas, month, yesterday, 20th century.

   My *birthday* will be on the first *Saturday* of next *month.*
   Christmas was *yesterday.*

7. **Activities**—driving, running, playing, sleeping, thinking, drinking.

   Drinking and *driving* do not mix well.
   The child's day was made up mostly of *running* and *playing.*

TECHNIQUE FOR IDENTIFYING NOUNS

Just remember that most nouns have two forms: **singular** (one) and **plural** (more than one).

one boy .... two boys
one salesperson .... a few salespeople
a book .... several books

(A few nouns, however, do not have plural forms—honesty, merchandise, running, etc.)

VERBS show or suggest action (doing) or existence (being). Every complete sentence includes at least one verb. The verb may be one action word, one existence word, or two to four words consisting of a main verb and one or more helping verbs:

1. ONE ACTION WORD: **type**, **read**, **have**, **think**, **go**, **dance**, and so on.

   (a) She *typed* her letter of application.
   (b) The personnel manager *read* his application carefully.
   (c) I *think* that I *have* a job for this young woman.
   (d) The salespeople *go* to San Francisco often.
   (e) He *danced* all night with the princess.

   [Notice that the forms of verbs change in spelling or in pronunciation to show the *time* of the action, as in Sentences (a), (b), and (e).]

2. ONE EXISTENCE WORD: An existence verb tells the "state of being" or permits *description* of the subject instead of telling what the subject "does."

BE: am, is, are, was, were, be, been, being.
SENSE: feel, sound, taste, smell, look.
ALSO: appear, become, seem.

(a) She *is* capable and efficient.
(b) I *am* happy about that.
(c) All the employees *were* on vacation last week.
(d) It *seemed* that he *was* in the pool all summer.
(e) The accounts payable manager *appears* efficient.

3. A GROUP OF WORDS MADE UP OF A MAIN VERB AND ONE OR MORE HELPING VERBS:

HELPING VERBS: BE group (am, is, are, was, were, be, been, being) . . . can, could . . . do, did, does . . . have, has, had . . . may, might, must . . . shall, should . . . will, would.

A helping verb expresses the approximate time (now, before, later, etc.) or a degree of emphasis of the main verb: **is reading** (now), **was typing** (before), **will have worked** (later), **has danced** (before), **have eaten** (before), **does sign** (emphasis), **did run** (emphasis+before), **were doing** (before).

(a) He *is reading* the application blank.
(b) She *was typing* all day.
(c) In June you *will have worked* for them for 25 years.
(d) He *has danced* with the princess before.
(e) She *does sign* all the letters herself.

HOW TO IDENTIFY VERBS

ING can be added to nearly all words that are verbs—seeing, having, eating, dancing, working, loving, cooperating, being, and the like. When these ING words express action or existence, they are verbs and are always preceded by a helping verb. See (a) and (b) below No. 3.

When ING words name activities, however, they are nouns and are not preceded by a helping verb. In the sentence "Typing is not difficult," typing is a noun because it names an activity.

REPLAY UNIT 1

(a) Underline the verbs in these sentences. When a **main** verb is combined with a **helping verb**, underline both.

(b) Circle the 54 nouns in these sentences. Use just one circle for names; for example, Ms. Adams.

NO. OF NOUNS IN EACH SENTENCE	EXAMPLE: Paul is helping Ms. Adams with her work.
1	1. Competent secretaries type accurately and rapidly.
2	2. Students frequently think about careers.
4	3. The manager and the accountant already have many reports on that subject.
3	4. Fred is the president of the company.
3	5. Ms. Adams was the manager of the store.
3	6. The clerks were typing the answers in the blanks.
3	7. The salesperson will have completed the work before Tuesday.
3	8. Martha should go to the conference in Kansas.
3	9. The wife of the manager had entered the lobby.
2	10. Mrs. Golden will ask for the letter.
2	11. These instructions have been provided for your convenience.
3	12. Ed knows a lawyer in that building.
3	13. Reuben Singer, the actor, will present a special concert.
2	14. Joe should mail the tickets.
3	15. The paper is in the box near the window.
2	16. Our bookkeeper will mail the invoice.
2	17. The Aldrich Company received the check.
3	18. Their auditor checked the books for accuracy.
4	19. Frankenstein built a factory in a city on the moon.
3	20. In this office the secretaries have new typewriters.

Match your answers with those on page 307.

READ UNIT 2

PRONOUNS

Pronouns are used as substitutes for nouns so as to avoid needless repetition of nouns. For example, the pronoun *she* could be a substitute for any of these nouns: Mary, girl, lady, typist, baby, president, woman, Ms. Anderson. These sample sentences show how pronouns act as noun substitutes.

NOUNS UNDERLINED	Miss Applebaum has finished Miss Applebaum's work.
PRONOUNS UNDERLINED	Miss Applebaum has finished her work. She has finished her work. She has finished it.
NOUNS UNDERLINED	Dick Tracy won the prize.
PRONOUNS UNDERLINED	He won it. Who won it? Somebody won it. Did anyone win this?

All you need to do in this section is learn to recognize a pronoun when you see one. (In Section Five, you'll find answers to questions you may have regarding when to use *who* and *whom* or *me* and *I*, etc.)

Pronouns that substitute for the names of specific kinds of people, animals, or things are grouped for convenience by these numbers: 1st, 2nd, 3rd. Just to make it easier to remember that these are always pronouns, they're shown below in these three groups.

FIRST PERSON	The person who is speaking or writing. **I, me, my, mine, myself**—singular; just me alone. **we, us, our, ours, ourselves**—plural; when you join me.
SECOND PERSON	The person to whom I'm speaking or writing. **you, your, yours**—singular or plural.* **yourself**—singular. **yourselves**—plural.
THIRD PERSON	The other person; the person or thing that I'm speaking or writing about. **he, him, his, himself**—singular, masculine. **she, her, hers, herself**—singular, feminine. **it, its, itself**—singular. **they, them, their, theirs, themselves**—plural.

Here are some other pronouns that are less specific or definite about whom or what they are substituting for:

*New Yorkers and Southerners, apparently frustrated by the absence of a plural 2nd person, devised their own—which are, of course, not Standard:
Youse—as in "Hey, youse guys, gimme da ball."
You'all—as in "Would you'all like some mint julep?"

who, whoever, whom, whomever, whose, this, that, these, those, someone, anyone, no one, everyone, somebody, anybody, nobody, everybody, something, anything, everything, nothing

> **NOTE**
> The words ending with *one, body,* and *thing* are called nouns in some dictionaries and pronouns in others. Since you will find more about them in the pronoun section of this book, it is helpful for you to identify them as pronouns.

Pronouns not only substitute for nouns, but they also substitute for other pronouns; for example—

Everyone should know *his* own telephone number. (Pronoun HIS is substituting for the pronoun EVERYONE.)

REPLAY UNIT 2

Part A

The directions are short and simple. If you read them carefully, you cannot make a mistake.

LIST THE **PRONOUNS** THAT SUBSTITUTE FOR:

1ST PERSON SINGULAR

1. The name of the person who is speaking or writing I me my mine myself

2ND PERSON

2. The name of the person you are speaking or writing to you your yours yourself yourselves

3RD PERSON FEMININE

3. The name of one female She her hers herself

3RD PERSON MASCULINE

4. The name of one male He His him himself

1ST PERSON PLURAL

5. Your name and my name together we us our ours ourselves

3RD PERSON SINGULAR

6. The name of one animal or thing It Its Itself

3RD PERSON PLURAL

7. The name of more than one animal, person, or thing They them their theirs themselves

Part B

Twenty-one other pronouns are listed on the previous page. LIST AS MANY AS YOU CAN REMEMBER

1. who
2. whom
3. whose
4. this
5. these
6. that
7. those
8. nobody
9. everybody
10. somebody
11. anything
12. everything
13. no one
14. everyone
15. everything
16. whoever
17. whomever
18. Someone
19. anyone
20. anybody
21. nothing

Answers are on page 308.

READ UNIT 3

ADJECTIVES

Adjectives *describe* or *explain nouns* or *pronouns.* As an aid to recognizing how adjectives function in sentences, four types of adjectives are shown below: descriptive, limiting, pointing, and articles.

You do not need to memorize the names of these four kinds of adjectives. They are presented only to help you identify adjectives by seeing how they are used.

DESCRIPTIVE ADJECTIVES

Many adjectives help the reader or listener picture the noun or pronoun. If you wish to communicate effectively about the noun "house," you might choose some *descriptive* words so that the listener or reader can picture the house you have in mind: brown and white, beautiful, shabby, two-story, luxurious, and so on. The clever selection of descriptive adjectives enables you to paint a picture for your reader with your typewriter.

Here are some adjectives that would be suitable for describing the nouns that are listed. You can probably think of additional descriptive words for each of these nouns.

Descriptive Adjectives	***Nouns***
abandoned, expensive, clean, old, blue	car
modern, electric, manual, broken, miniature	typewriter
delicious, nutritious, gourmet, hot, Mexican	food
Japanese, stingy, pleasant, short, honest	man

SAMPLE SENTENCES USING DESCRIPTIVE ADJECTIVES

1. The blue car is old but clean.
   (*Blue*, *old*, and *clean* all describe the noun *car*.)
2. Did you see the miniature typewriter on the shelf?
   (*Miniature* describes the noun *typewriter*).
3. Gourmet food is served in that restaurant.
   (*Gourmet* describes the noun *food*.)
4. That man is honest and handsome.
   (*Honest* and *handsome* both describe the noun *man*.)
5. He is honest and handsome.
   (*Honest* and *handsome* both describe the pronoun *he*.)

LIMITING ADJECTIVES

Words that "limit" nouns in the sense of quantity are also adjectives. Here are some adjectives suitable for "limiting" the nouns listed.

Limiting Adjectives	***Nouns***
more, several, many	typewriters
fifty, enough, few, some	workers
no, most, any, 22	ideas

SAMPLE SENTENCES USING LIMITING ADJECTIVES

1. Several typewriters were stolen last month.
   (The adjective *several* is used to "limit" the noun *typewriters*.)
2. There aren't enough workers to do the job properly.
   (The adjective *enough* limits the noun *workers*.)
3. He submitted 22 ideas for the project.
   (The adjective *22* limits the noun *ideas*.)

POINTING ADJECTIVES

There are four words that are sometimes used to "point" at nouns—just as one might do with an outstretched finger. When these words have the "pointing" function in a sentence, they are adjectives.

Pointing Adjectives for Use with Singular Noun	*Noun*	*Pointing Adjectives for Use with Plural Noun*	*Noun*
this that	book machine	these those	books machines

I just heard an alert student protest, "Hold on a minute, Ms. Smith. In Unit 2 you told us THIS, THAT, THESE, and THOSE are pronouns. Now you're telling us they're adjectives!"

THIS, THAT, THESE and THOSE are *adjectives* when used before a noun:

This book **is the one I want to read.**

THIS, THAT, THESE, and THOSE are *pronouns* when the noun is left out:

This **is the one I want to read.**

ARTICLES

Another type of adjective is known as an "article." There are only three articles—*a, an, the.* (A review of when to use *a* or *an* appears in Section 7.)

SAMPLE SENTENCE USING ARTICLES

An apple *a* day keeps *the* doctor away.

HOW TO IDENTIFY ADJECTIVES
Any word that describes, explains, or restricts a noun or a pronoun is an ADJECTIVE. Adjectives usually appear before nouns

classical music, *red* dresses, *hot* potatoes

or after existence verbs.

He is *handsome*. Ms. Murcar is *rich*. It is *good*.

REPLAY UNIT 3

Part A

The following paragraph is part of a sales letter sent to many people for the purpose of bringing business to a resort hotel. Most of the adjectives in this paragraph are the "descriptive" kind and are used to "paint a picture," while others are the type known as "articles."

Surveys show that you are a person who should be interested in visiting an exclusive resort community, located on the beautiful Pacific, where the crystal blue waters meet the white sandy beaches. You will enjoy exquisite guest rooms of Spanish decor, gourmet dining, and a heated pool. The view is serene, smogless, and sunny.

The numbers in the *Adjectives* column indicate how many adjectives in the paragraph above describe the noun in the next column. The first one is an example; that is, just one (1) adjective, *a*, describes or explains the noun *person*. On the next line fill in the three adjectives that describe the noun *community*. Continue until you have completed all the blanks.

	ADJECTIVES	NOUNS
Example	(1) a	person (1st line)
	(3)	community (2d line)
	(2)	Pacific (3d line)
	(3)	waters (3d line)
	(3)	beaches (3d line)
	(2)	rooms (4th line)
	(1)	decor (4th line)
	(1)	dining (4th line)
	(2)	pool (5th line)
	(4)	view (5th line)

Part B

These sentences contain eight limiting and pointing adjectives. Underline each one.

1. These offices don't have any carpeting.
2. They have many windows, however.
3. This office is mine; that is yours.
4. There are five departments on some floors.
5. Several managers have quit this year.

Part C

On the line below write five adjectives that could be used to describe the noun OFFICE.

Check answers on page 308.

Now turn to page 263 and take the POP QUIZ on Units 1–3.

READ UNIT 4

ADVERBS

Like adjectives, adverbs also describe or explain. In Read Unit 3 you learned that *adjectives* describe or explain *nouns* or *pronouns. Adverbs*, however, *describe* or *explain*:

Verbs
Adjectives
Other Adverbs

Adverbs do *not* answer the questions *Who*? or *What*? about the words they describe. The questions they do answer are *When*? *Why*? *How*? *How much*? *Where*? *To what extent*?

Barbara is going immediately. (When?)
Michael is going away. (Where?)
Robert is going quickly. (How?)
The gum costs very little. (How much?)

HOW TO IDENTIFY ADVERBS

(a) Any word that describes or explains a verb, an adjective, or an adverb is an ADVERB.
(b) Many adverbs (but not all) are formed by adding "ly" to an adjective.

SOME ADJECTIVES CONVERTED TO ADVERBS

quick	quickly
quiet	quietly
exceptional	exceptionally
intelligent	intelligently
attractive	attractively
final	finally
real	really

SOME WORDS OFTEN USED AS ADVERBS THAT DON'T END IN "ly"

almost, even, more, much, never, not, so, too, very, well

USE AN ADVERB TO DESCRIBE A VERB

Study these sentences to see how the underlined adverbs describe the verbs (labeled v). Say the verb followed by **HOW?** The adverb answers the question.

1. She types (v) carefully.
2. She types (v) well.

3. She works (v) quickly and efficiently.

4. He prints (v) rapidly and accurately.

USE AN ADVERB TO DESCRIBE AN ADJECTIVE

Study these sentences to see how the underlined adverbs describe the adjectives (labeled adj). The adjectives describe the nouns (labeled n). Say **HOW** before the adjective. The adverb answers the question.

1. The book (n) is extremely expensive (adj)
2. The wallpaper (n) looks very attractive (adj).
3. The most difficult (adj) courses (n) are given in the morning
4. Marjorie (n) is too intelligent (adj) to have said that.

USE AN ADVERB TO DESCRIBE ANOTHER ADVERB

Study these sentences to see how the underlined adverbs describe other adverbs (labeled adv). The explanations below each sample sentence should be read carefully as a help in identifying the various parts of speech.

1. They are working exceptionally hard (adv).
   Adverb HARD describes verb ARE WORKING.
   Adverb EXCEPTIONALLY describes adverb HARD.
2. Mr. Johnson works so efficiently (adv) that he almost never (adv) makes a mistake.
   Adverb SO describes adverb EFFICIENTLY.
   Adverb EFFICIENTLY describes verb WORKS.
   Adverb ALMOST describes adverb NEVER.
   Adverb NEVER describes verb MAKES.
3. Our company manufactures tools very cheaply (adv).
   Adverb VERY describes adverb CHEAPLY.
   Adverb CHEAPLY describes verb MANUFACTURES.

REPLAY UNIT 4

Insert an adverb in each blank. The adverb you select will describe or explain the verb, adjective, or adverb that is underlined and will answer the question When? Why? Where? How? How much? or To what extent? To be sure you select an adverb, refer to the box in Read Unit 4. Use only one word for each blank.

Example Ms. White dresses (v) ___elegantly___. (or well, beautifully, casually, expensively, etc.)

1. He was discharged [v] ________.
2. The children in that class read [v] ________.
3. Do you think that Ms. Lopez drives [v] ________?
4. You did [v] not add [v] the figures ________.
5. That machine works ________ well [adv.].
6. Mrs. Warner is an ________ intelligent [adj] person.
7. The books are ________ expensive [adj].
8. They are working ________ hard [adv].
9. I can't keep up with him because he runs ________ fast [adv] for me.
10. They should ________ use [v] good quality paper for that job.
11. When you go [v] ________ you will find that the front door was left open.
12. A new adding machine would cost [v] ________ than a used one.
13. He ________ attends [v] the weekly meetings.
14. I think Clark Gable was ________ more [adv] attractive than your boyfriend.
15. Nutrition is one of the ________ important [adj] parts of this program.
16. This ________ designed [adj] home was described in the *Architectural Digest.*
17. We ________ gave [v] the old duplicator away.
18. Word processing equipment is ________ expensive [adj], but it results in long-range economies for the business.
19. We didn't believe Delilah when she said she had [v] ________ cut [v] Samson's hair.
20. The story is ________ true [adj].

See possible answers on page 309.

READ UNIT 5

LITTLE WORDS

CONJUNCTIONS

Conjunctions *join* two words or two groups of words (such as phrases or sentences). The most commonly used conjunctions and the easiest to recognize are:

and but or nor

They are often used along with a comma to join two complete thoughts or sentences into one longer sentence.

> Mildred Sweet is a display rack saleswoman, **and** Melanie Sweet is a medical student.
>
> Kathy Allen is a talented artist, **but** she prefers to groom poodles.

These conjunctions are used in other ways as well. In the following sentences, they do not join two complete thoughts. That's why commas are not used before them.

> We can ship the merchandise either today **or** tomorrow.
> He will neither see you **nor** speak with you.
> Ms. Jackson seems to work hard **but** doesn't accomplish much.
> The Beatles **and** the Beach Boys were the most famous rock groups of the 1960s.

The conjunctions **and**, **but**, **or**, **nor**, and **for** are called "coordinate" conjunctions. They are used to join two equal things such as two complete thoughts, two nouns, two verbs, and the like.

Some other words often used as conjunctions are:

if, as, when, because, since, although,
before, until, unless, after,
though, even though, while,
so that, whenever, than

These are called "dependent" or "subordinate" conjunctions. They are used before a noun or pronoun followed by a verb, as shown below:

> The new Bond's branch will be successful if Mickey manages it.
>
> Our Apple cannot be used for preparing the report unless it is repaired this week.

Often the normal order of a sentence is reversed; then the conjunction might be the **first** word of the sentence. If the dependent conjunction does come first, use a comma between the parts of the sentence as shown in the two sentences that follow:

> If Mickey manages the new Bond's branch, it will be successful.
> Unless it is repaired this week, our Apple cannot be used for preparing the report.

After reading Unit 12, you'll know more about dependent conjunctions. For now, just identify the words centered above as conjunctions when they come before a noun or pronoun **followed by a verb.**

PREPOSITIONS

A preposition is a word that comes before a noun or pronoun to form a prepositional phrase—a word group that does **not** contain a verb.

A preposition cannot operate by itself. It is meaningless unless it has a noun or a pronoun that comes right after it or soon after it. This noun or pronoun is called the OBJECT of the PREPOSITION. The preposition and its object plus any words that come between them are known as a PREPOSITIONAL PHRASE.

A prepositional phrase is a group of related words that begins with a preposition and ends with the noun or pronoun that is its object. Here is a list of words often used as prepositions:

about	behind	from	over
above	below	in	since
across	beneath	inside	through
after	beside	into	to
against	between	like	toward
along	by	near	under
among	during	of	up
around	except	off	upon
at	for*	on	with

memorize

Since only the preposition and its object are essential for making a prepositional phrase, the phrase might consist of just two words like these:

Preposition Plus Pronoun Object	***Preposition Plus Noun Object***
to me	by Mary
of him	near London
with you	except caterpillars

Sometimes other words (nouns and pronouns that show ownership, adjectives, and adverbs) are used in the middle. However, the preposition is always first and the noun or pronoun last. In these phrases, the interrupting words are underlined:

under the antique table	below the newly painted roof
with my adorable little sister	through the air
in the decrepit plane	across the street
from Mary's window	in the leather briefcase

SAMPLE SENTENCES USING PREPOSITIONAL PHRASES (phrases underlined)

Notice that the preposition shows the relationship between the object of the preposition and the rest of the sentence.

I hid / under the antique table / with my adorable little sister. /
I will go / with you / in the decrepit plane / to a village / near London. /
The burglars, who were quite clever, hid the diamonds / in the leather briefcase. /
I saw the newly painted roof / from Mary's window. /

For is a conjunction when it means "because." Otherwise, it's a preposition.

REPLAY UNIT 5

Part A

List some commonly used conjunctions that are mentioned in Read Unit 5.

1. up
2. to
3. about
4. above
5. across
6. below
7. for
8. on
9. off
10. near

like

Part B

Almost any word that makes sense in the sentence below is used as a preposition. The prepositions you write in the blanks will show the relationship between the verb **could go** and the noun **mountain**. (To be sure you have all correct answers, just be sure you don't choose a **verb**, such as "climb," for any of the blanks.)

They could go __________ the mountain.

__________ __________

__________ __________

__________ __________

__________ __________

Part C

Insert a suitable conjunction in the blank.

1. Neither Susan __nor__ Jeffrey wishes to leave the New York area.
2. Ivy has been a commercial artist for an advertising agency, __but__ she prefers to do landscape painting.
3. No one was there to meet them __when__ they arrived.
4. Earl Hines was raised in Pittsburgh, __and__ that's where he started his jazz career.
5. "__although__ I used my formal piano training as a foundation, I've always tried to be original," said Hines.

Part D

Underline the prepositional phrases.

1. He will be in touch with you during the first part of next week. (4 phrases)
2. Advertising has a substantial influence on the behavior of people and on their lifestyles. (3 phrases)
3. There are many types of media. (1 phrase)
4. Media for advertising are rated in several ways. (2 phrases)
5. Radio advertisers are experts at producing spot announcements for their customers. (2 phrases)

Please check your answers on page 309.

READ UNIT 6

VERSATILE TOOLS

Some tools are quite versatile because they are used for more than one kind of job. So it is with English words. The seven parts of speech briefly reviewed in this section can be identified with certainty only when you observe how words are *used* in sentences. That is, a great many words may be used as more than one part of speech. You can't really be sure of the part of speech of many words unless you see them in a particular sentence.

The idea is simple: Just remember that the part of speech is determined by what the word *does* for the sentence. Consider the word "dancing." It looks like a verb, doesn't it? After all, it represents *action*. It sounds like a verb too—because it ends in *ing*. As a matter of fact, it *is* a verb . . . sometimes.

He *is dancing* with a princess.

IS DANCING is the complete verb of that sentence—made up of the helping verb IS and the main verb DANCING. The *ing* form is a verb when a helping verb precedes it.

THEREFORE

Dancing is enjoyable.

DANCING is a noun because it is the *name* of an activity; it is the subject of the sentence, and IS is the verb.

AND

The child attends *dancing* school.

DANCING is an adjective because it describes the school; ATTENDS is the verb.

When there's no helping verb, the *ing* word is a noun or an adjective.

How about the word "brown"? What part of speech do you think it is? Since it's a color, it must be a describing word—an adjective. That's right . . . sometimes.

Brown carpeting was installed in the treasurer's office.

BROWN is an adjective describing the noun CARPETING.

BUT

Mr. *Brown* is the new treasurer.

BROWN is a noun because it names a person.

OR

Brown is my favorite color.

BROWN is a noun because it's functioning only as the **name** of a color and is the subject of the sentence. (More about subjects in Unit 8.)

AND

First, I *brown* the onions.

BROWN is now the action word of the sentence—a verb.

REPLAY UNIT 6

Part A

The word FISH is used as a noun, an adjective, or a verb. What part of speech is FISH in each of these sentences?

1. He will fish when he goes on vacation. ____________
2. We'll eat a fish dinner. ____________
3. Fish is my favorite food. ____________

Part B

The word TRADE may also be a noun, an adjective, or a verb. What part of speech is TRADE in each of these sentences?

1. That was not a fair trade. ____________
2. I will not trade that stock. ____________
3. He reads several trade papers. ____________

Part C

The word TYPING is another word that readily changes its part of speech. Write a sentence using TYPING as each part listed.

1. (verb) ______________________________
2. (noun) ______________________________
3. (adjective) ______________________________

Part D

Write two sentences using STEP, first as a verb and then as a noun.

1. (verb) ______________________________
2. (noun) ______________________________

Part E

1. The words CITY and COUNTRY are what part of speech in the following sentence? The researcher plans to compare city schools with country schools. ______________
2. Write a sentence in which city and country are nouns.

   ____

   ____

The answers to this Replay are on page 309.
For POP QUIZ, turn to page 264.

READ UNIT 7

COURTEOUS COMPLAINT

This courteous complaint letter illustrates Standard English for business. The sentence structure, grammar, punctuation, and spelling are correct; the word choices are suitable; and the content is effectively arranged. As you read this letter and others in this book, you will acquire the "knack" of recognizing good business communication.

Please also study the various parts of this personal business letter. Notice that the writer's home address is typed above the date, and the writer's name is typed below the signature. The other parts of a personal business letter are the inside address, salutation, body, complimentary close, and in this case, a "copy" notation indicating that a copy was sent to someone other than the addressee. You'll find excerpts from business letters and complete business letters in several sections of the book.

All the parts of speech reviewed in this section are, of course, included in this letter: nouns, verbs, pronouns, adjectives, adverbs, prepositions, conjunctions.

REPLAY UNIT 7

1. What part of speech are these words taken from the first paragraph of the letter?

   (a) we ________________ (e) prompt ________________

   (b) wagon ________________ (f) also ________________

   (c) purchased ________________ (g) appreciate ________________

   (d) on ________________

2. What are the two verbs in the first line of the second paragraph?

   (a) ________________ (b) ________________

3. What are the two pronouns in the fourth line of the second paragraph?

   (a) ________________ (b) ________________

4. What part of speech is the last word of the second paragraph?

   ____

5. What parts of speech are these words from the last paragraph?

   (a) has been ________________ (e) us ________________

   (b) some ________________ (f) your ________________

   (c) error ________________ (g) dealer ________________

   (d) or ________________

Answers are on page 309.

Writer's Address

30711 Fairplay
Wilmington, DE 19809
July 1, 198-

Date

Owner Relations Department
Chevrolet Central Office
Chevrolet Motor Division
Detroit, MI 42842

Inside Address

Ladies and Gentlemen:

Salutation

Body

We are very much pleased with our new Chevrolet station wagon, which we purchased on June 21. The dealer, Morris Owens Chevrolet, has extended us courtesy and prompt service, which we also appreciate.

We believe that this dealer and General Motors conduct business fairly and honestly. As we all know, it is common practice for a manufacturer or supplier to replace or repair its product without charge if it had a defect at the time of purchase. Yesterday we paid $82.20 for repair of a defect that was in existence when we bought the car. At the time of purchase, although unknown to us, the wheels needed balancing and weights.

We note the exclusion provisions of the warranty, which read, "Normal Maintenance services (such as. . . wheel balancing)" are not covered by the warranty. A condition existing at the time of purchase cannot of course be considered "Normal Maintenance."

We assume, therefore, that there has been some error or misunderstanding. Would you please send us your check (or instruct your dealer to do so) for $82.20.

Complimentary Close

Sincerely yours,

Writer's Signature

Leila R. Smith

Leila R. Smith

Writer's Name

Copy Notation

c: Morris Owens Chevrolet Service Department

Checkpoint

The "tools of the trade" (the seven parts of speech) were presented briefly in this section. The information provided is not sufficient for identifying the part of speech of every word or word group in every sentence since this ability is not of much value to most students.

You will be able, however, to recognize parts of speech so that you can master the English principles presented in this book. Learning the tools of the English trade will make the rest of the work that follows much easier. That's why if you devote some extra concentration to this section, you'll have a smooth sail through the rest of the sections. Write a check in the blank after you have learned the definition.

____ NOUN—Names something or somebody.
To decide whether a word is a noun, see if it sounds right with a verb after it. For example,

n v
Beauty is in the eye of the beholder
n v
Sports are an imporıant part of physical education.

____ PRONOUN—substitutes for a noun.
To decide whether a word is a pronoun, see if it can substitute for a noun. For example,

pro pro n n
Give that to me. = Give the tape to Ms. Smith.
pro pro pro n n
It injured its paw in our yard. = The squirrel injured the squirrel's
n n
paw in Joe and Ann's yard.

____ VERB—Expresses action or existence.
To decide whether a word is a verb, see if it sounds right with a noun before it. For example,

n v
Accountants have many responsibilities.
n v
Technicians were paid more.

____ ADJECTIVE—Describes, points to, or limits a noun or pronoun. An "article" (a, an, the) is a form of adjective.

____ ADVERB—Describes a verb, an adjective, or another adverb. Adverbs often (but not always) end with "ly."

____ CONJUNCTION—Joins two words or groups of words. The easiest to recognize conjunctions are AND, BUT, OR, and NOR. Dependent conjunctions precede a noun or pronoun + verb combination. Examples are WHEN, AS, IF, and ALTHOUGH.

____ PREPOSITION—Introduces a prepositional phrase. A prepositional phrase does not contain a verb. The last word of a prepositional phrase is a noun or a pronoun known as the object of the preposition.

For additional details regarding the parts of speech, see the Grammar Supplement, page 294.

Special Assignment

- The ability to present a clear explanation or description is vital to successful business, professional, and technical communication. This assignment will provide practice in expressing yourself clearly.

  You and your partner on a TV game show have a chance to win a new Mercedes and a trip to Europe. All you have to do is describe the object (without naming it) given to you by the master of ceremonies. Your partner must name the object, which he cannot see. Your description should be so vivid that your partner gets it right on the first try. You'll probably want to tell him facts such as size, weight, shape, texture, color, or whatever you feel would be helpful. The rules require complete sentences and a maximum of fifty words. (Your instructor will give you the name of the object.)

- Compose sentences using each of these words as an adjective, a verb, and a noun: running, letter, and radio.

The assignments should be completed on _______________ (date).

> **WEBSTER'S DICTIONARY**
> Several good dictionaries carry the name of Webster as part of the title; however, the name Webster does not in any way ensure the quality of a dictionary. "Webster" cannot be copyrighted; thus anyone can compile a reference book and give it that name. (Noah Webster was a 19th-century lexicographer who compiled THE AMERICAN DICTIONARY, an outstanding language reference book of its time.)

Proofreading for Careers

Everyone makes mistakes; even I have made one or two in my lifetime. But for success on the job, why advertise mistakes! Instead PROOFREAD, PROOFREAD, PROOFREAD.*

Proofreading

- tests alertness and knowledge.
- is not like other kinds of reading.
- is slower than other types of reading.

This classified advertisement appeared in the *Kingsport* (Tennessee) *News*. Can you improve on the proofreading?

> Telephone receptionist for doctors' office. Duties are to relay massages between patients and doctors.

*Special symbols are often used for proofreading. Typists interpret these symbols and make corrections accordingly. These symbols are explained under the heading "Proofreaders' Marks" in college dictionaries, typing books, and reference manuals.

In this *Los Angeles Times* headline, medical history was made. Proofread and correct it:

> TACKLE TO UNDERGO KNEE SURGERY FOR THIRD TIME TODAY.

From this answer on a science student's test, biology history was made. Proofread it for "sense" and make the correction:

> An example of animal breeding is the farmer who mated a bull that gave a great deal of milk with a bull with good meat.

Proofreading is much more than checking for spelling and typographical errors. Don't let anything get by your desk that doesn't make sense to you. Verify your corrections with the key or with your instructor. What's your score in proofreading? ________ A=Alert
M=Mediocre
U=Ugh

Practice Quiz

Part A

Fill in the blank with the name of a part of speech.

1. To name a person or a thing, use a/an NOUN. (Unit 1)
2. To describe a noun or a pronoun, use a/an adjective. (Unit 3)
3. To describe a verb, adjective, or adverb, use a/an adverb. (Unit 4)
4. To substitute for a noun, use a/an PRONOUN. (Unit 2)
5. To join two words or groups of words, use a/an conjunction. (Unit 5)
6. To begin a prepositional phrase, use a/an preposition. (Unit 5)
7. To express action or existence, use a/an verb. (Unit 1)

Part B

After reading the sentence, fill in the part of speech of each word.

> WHILE SARALA DINED LEISURELY, SHE THOUGHT ABOUT SKIING WITH SEAN IN THE LOFTY MOUNTAINS OF VERMONT AND NEW HAMPSHIRE.

8. While conjunction. (Unit 5)
9. Sarala NOUN (Unit 1)
10. dined verb (Unit 1)
11. leisurely adverb (Unit 4)
12. she PRONOUN (Unit 2)
13. thought verb (Unit 1)

14. about ___Preposition___ (Unit 5)

15. skiing ___NOUN___ (Unit 1)

16. with ___Preposition___ (Unit 5)

17. Sean ___NOUN___ (Unit 1)

18. in ___Preposition___ (Unit 5)

19. the ___article / adjective___ (Unit 3)

20. lofty ___adjective___ (Unit 3)

21. mountains ___NOUN___ (Unit 1)

22. of ___Preposition___ (Unit 5)

23. Vermont ___NOUN___ (Unit 1)

24. and ___conjucntion___ (Unit 5)

25. New Hampshire ___NOUN___ (Unit 1)

PRACTICE QUIZ GRADE: ______

No. Right	Grade
25	A+
24	A
23	A−
22	B+
21	B
20	B−
19	C+
18	C
17	C−
16	D+
15	D
14	D−

SECTION 2

SECRET LIFE OF A SENTENCE REVEALED

After Completing Section Two, You Will:

- *Identify a fragment, run-on, or comma splice and correct it.*
- *Construct complete sentences and end them at the appropriate place.*

We have all learned to derive meaning and understanding from certain combinations of letters. For example, when we combine the following letters in the order shown, we can pronounce a word and understand it:

S-P-E-L-L

We can't, however, be sure of whether SPELL refers to magic or to the arrangement of letters to form a word. (Perhaps the latter is also magic.)

Thus, to increase our understanding of words, we arrange them into groups. Read the three sample word groups below:

1. **A when not word spell how to you sure are.**
   Even though the group of words begins with a capital letter and ends with a period, you know it isn't a sentence. You don't receive a sensible message because you're not accustomed to this kind of word arrangement.
2. **When you are not sure how to spell a word.**
   Now you receive a sensible message because you are accustomed to this kind of word arrangement. But you probably have a feeling of incompleteness; you sense that something is missing. Even though you can interpret the group of words, you are waiting for completion of the thought.
3. **When you are not sure how to spell a word, you should look it up in the dictionary.**
   At last the familiar signals—capital letter and period—have enclosed a *complete* thought that *can* be understood.

Clearness is a very important goal in business writing, and complete sentences are the first step toward that goal. Good business writers use complete sentences in memos, letters, and reports. Incomplete sentences are called FRAGMENTS.

The opposite of a **fragment** (or incomplete sentence) is the RUN-ON sentence. This is what happens when writers forget to stop. They go on and on and don't insert a period until the end of two or even several complete thoughts:

Run-on Mr. Lemos hired a new secretary she can take dictation at 120 words a minute.

Some writers remember to stop between two complete thoughts, but not firmly enough. They are guilty of the COMMA SPLICE, which is a run-on sentence with a **comma** between the complete thoughts.

Comma Splice Mr. Lemos hired a new secretary, she can take dictation at 120 words a minute.

When fragments, run-ons, or comma splices appear in business writing, the reader is likely to be distracted from the important message contained in the letter or report. In addition, the reader may silently question the educational level or thinking ability of the writer.

You will be confident of your ability to avoid fragments, run-ons, and comma splices in your own writing when you have completed the READ AND REPLAY units that follow.

READ UNIT 8

PSYCHOLOGY AND SENTENCE CONSTRUCTION

Just as you and I have basic needs, a sentence also has basic "needs." When our needs are not met, we feel a lack of completeness; we are "frustrated." When the needs of a sentence are not met, it is "incomplete"; it is a FRAGMENT. To be "complete," a sentence reveals three characteristics:

Identity Action Independence

IDENTITY

A sentence must include at least one word that identifies **who** or **what** the sentence is about. This part of the sentence is called the SUBJECT. The subject is always a noun or a pronoun.

SAMPLE SUBJECTS

The underlined words are nouns or pronouns that identify **who** or **what** the sentence is about.

1. Ms. Ramirez is an expert secretary. (who)
2. The new treasurer stole our money. (who)
3. My coat is hanging in the closet. (what)
4. It is torn but can be repaired. (what)
5. She takes dictation at 120 words a minute. (who)
6. Did he steal your fur coat also? (who)
7. The windows and doors had been left open. (what)
8. The typewriter isn't in very good condition. (what)
9. The manager ordered a new typewriter for him. (who)
10. The desk is just the right size. (what)

ACTION

A sentence must include at least one word that tells what the subject **does** or **is** or **has**. This part of the sentence is called the VERB.

The verbs below are underlined. They tell what the subject *does*, *is*, or *has*. The subject (who or what the sentence is about) is shown with an asterisk.

1. Ms. Ramirez* types a letter efficiently.
2. She* also knows all about various filing systems.
3. My horse* ate the hay.
4. Then he* trotted around the race course.
5. The treasurer* has caviar and champagne.
6. He* flew to South America with the money.
7. Embezzlement* is a serious crime.
8. Was your coat* in the closet?
9. The horse* seemed sad and lonely.
10. I* worked at the typewriter all day.

REPLAY UNIT 8

Part A

(a) The verb tells what the subject **does** or **is** or **has**. Write the verb of each sentence in the blank.

(b) The subject tells **who** or **what** is doing something; it tells who or what the sentence is about. Write the subject of each sentence in the blank.

I suggest that you find the verb first. Then ask *Who*? or *What*? before saying the verb aloud. After identifying the verb in the sample sentence, say, "Who or what *is*?" Then answer, "*secretary* is."

		SUBJECT	VERB
Sample	The new secretary is in the office now.	secretary	is
1.	Mr. Johnson was the president.	Mr. Johnson	was
2.	You introduced her to the staff yesterday.	you	introduced
3.	The typewriters in this office are electronic.	typewriters	are
4.	This assignment seems easy.	assignment	seems
5.	The books were on the shelf.	books	were
6.	Mr. Johnson wants the report.	Mr. Johnson	wants
7.	The new members of the team have uniforms now.	members	have
8.	We walk home every day.	we	walk
9.	The secretary types the report accurately.	Secretary	types
10.	The supervisor found five errors.	Supervisor	found

Notice that the object of a preposition can't also be the subject of a verb. A prepositional phrase does not contain a subject and verb. A word group that **does** contain a subject and verb is a CLAUSE.

Part B

Part A, which you completed above, contains ten **complete** sentences. Part B below contains ten **incomplete** sentences—FRAGMENTS. They are fragments because either the subject or the verb is missing. Write *S* in the blank if the subject is missing or *V* if the verb is missing.

		MISSING
Sample	Our new secretary, Ms. Ramirez.	V
	1. The former treasurer of this company.	V
	2. Runs for exercise every day	S
	3. Usually eats hay every morning.	S
	4. Costs a great deal of money.	S
	5. The big black race horse in my stable.	V
	6. Was caught stealing the sugar.	S
	7. A mink coat with a sable collar and a silk lining.	V
	8. Is a very good typist.	S
	9. The little old lady in tennis shoes.	V
	10. Our experienced and competent bookkeeper.	V

Part C

Use the ten fragments in Part B to compose five complete sentences. Combine them in any way you wish as long as you create sentences that include both subject and verb.

1 & 2 1. the former Treasurer of this company, runs for exercise every day.

5 & 3 2. The big black race horse in my stable, usually eats hay every morning.

7 & 4 3. a mink coat with a sable collar and a silk lining, cost a great deal of money.

8 & 6 4. the little old lady in tennis shoes, was caught stealing the sugar.

10 & 8 5. Our experienced and competent bookkeeper, is a very good Typist.

The answers are on page 310.

READ UNIT 9

THE CLINGING VINE

An old-fashioned girl who clung to her boyfriend and could not do anything without him was known as a clinging vine. Although there are not many women like that anymore, the idea brings us to the third (and final) requirement of the complete sentence—its INDEPENDENCE.

A sentence needs to be able to "stand alone"—that is, it must be *independent* in order to be *complete*. When a group of words cannot "stand alone," it is *dependent*. This means that it must lean on another group of words in order to form a complete sentence. A group of words may have both subject and verb but still not express a complete thought. Such a word group is sometimes called a *dependent clause*. A dependent clause that begins with a capital letter and ends with a period is a FRAGMENT; it is not a sentence.

INDEPENDENT (SENTENCES)	DEPENDENT (FRAGMENTS)
(a) Ms. Ramirez types a letter	(a) If Ms. Ramirez types a letter.
(b) The treasurer ran away.	(b) Because the treasurer ran away.
(c) Her fur coat cost $500.	(c) Since her fur coat cost $500.
(d) My horse eats hay.	(d) As my horse eats hay.

Although (a), (b), (c), and (d) in both columns include subjects and verbs, only the ones on the left (the independents) are complete sentences.

The four *dependents* on the right are FRAGMENTS because the first word (a dependent conjunction) causes the group of words to be "dependent." When a dependent doesn't have anything to lean on or to "depend on," it lacks completeness. Therefore, it is not a complete sentence even though it has a subject and a verb.

The dependents* (or fragments) can be completed by adding an *independent* group of words—that is, a group of words capable of standing alone. A group of words that expresses a complete thought is often called an **independent clause**.

The underlined expressions below are **independents**:

(a) If Ms. Ramirez types a letter, she will surprise her supervisor.
(b) They can't build the new factory because the treasurer ran away.
(c) Since her fur coat cost $500, she is very careful with it.
(d) We don't need any oats, as my horse eats hay.

Each underlined group of words has identity (subject), action (verb), and independence—the three requirements of a sentence. That's why each underlined group could "stand alone" as a complete sentence. The word group not underlined is dependent and would be a fragment if it were not attached to an independent. Some of the dependent conjunctions that make a word group dependent are:

after, although, as, because, before, even though, if, provided, since, so that, that, unless, until, when, where, whether, which, while

*Also known as subordinate clauses.

REPLAY UNIT 9

Part A

The word groups below include both fragments and complete sentences. Notice how the fragments leave you with a feeling of incompleteness. You feel that a complete thought has not been expressed; something is missing. Write *F* in the blanks next to the fragments and *C* next to the complete sentences.

______ 1. If you do a good job on this project.

______ 2. You will get a raise in salary after the first of the month.

______ 3. Although most corporations use the services of an auditor to examine the books.

______ 4. Some errors may never be found.

______ 5. Synthetic fabrics are generally easy to launder.

______ 6. Even though some stains never come out.

______ 7. Before she became the manager.

______ 8. Important papers were difficult to locate.

______ 9. She will start to work for the new treasurer.

______ 10. Now that Ms. Ramirez has completed the training period.

Part B

Use the ten items in Part A to compose five proper sentences. Combine each fragment with one of the complete sentences.

1. ______________________________

2. ______________________________

3. ______________________________

4. ______________________________

5. ______________________________

The answers to this Replay are on page 310.

READ UNIT 10

FRAGMENTS, ANYONE?

Here are samples of another type of fragment to guard against. These are also **dependents**, but their construction differs from those of Unit 9.

1. MR. HENRY, WHILE JOGGING AROUND THE BLOCK. (FRAGMENT)

In order to feel that you have read a complete sentence, you want to know what happened to Mr. Henry. Here are two of the ways the fragment can be made into a complete sentence:

(a) Mr. Henry, while jogging around the block, sprained his ankle.

or

(b) Mr. Henry was jogging around the block.

By changing "while" to "was," you construct a complete sentence. A sentence that follows might include the information about his sprained ankle.

2. MR. LOPEZ, WHO IS THE PRESIDENT OF THE COMPANY. (FRAGMENT)

For a feeling of completion, you want to know something else about Mr. Lopez. Here are two of the ways that the fragment can be corrected:

(a) Mr. Lopez, who is the president of the company, saw Mr. Henry fall.

or

(b) Mr. Lopez is the president of the company.

By omitting the word "who," you construct a complete sentence. Other information about Mr. Henry's fall can be part of a new sentence.

3. MR. HENRY'S OFFICE, WHICH IS CLOSED TODAY BECAUSE OF HIS FALL. (FRAGMENT)

Notice that to feel you have read a complete sentence, you want to know something more about Mr. Henry's office. The fragment can be corrected in several ways:

(a) Mr. Henry's office, which is closed today because of his fall, will be open all of next week.

or

(b) Mr. Henry's office is closed today because of his fall.

By omitting "which," you change the fragment into a complete sentence.

REPLAY UNIT 10

Part A

Write *F* for fragment or *S* for sentence in the blanks below.

F 1. The man whom we met yesterday.

F 2. The personnel department, which is on the third floor.

F 3. Mr. McCarthy, feeling that he was right.

S 4. Mr. McCarthy felt that he was right.

S 5. Mr. McCarthy, feeling that he was right, continued to argue the point.

F 6. The staff, having been on vacation last week.

S 7. The staff was on vacation last week.

F 8. The team that won all the games last year.

S 9. That team won all the games last year.

S 10. The team that won all the games last year is losing today.

Part B

Five of the lines above are fragments. Write the line number in the blank below; then correct the fragment so that it becomes a complete sentence. (Make up any necessary information that will complete the sentence.)

Sample 1. The man whom we met yesterday is the treasurer who ran away.

2. The personnel department, which is on the third floor, will move out in 3 month.

3. Mr. McCarthy, feeling that he was right complaint to the boss.

6. the Staff, having been on vacation last week went back to work today.

8. the S.F. 49ers is the team that won all the games last year is loosing Today.

Please check your answers on page 310.

Now review Units 8–10; then take the Pop Quiz on page 265.

READ UNIT 11

CAPITAL PUNISHMENT

RIDDLE: What kind of sentence requires capital punishment?

HINT NO. 1
Read this sentence taken from a college student's essay:

> Cindi is a carefree and somewhat irresponsible girl, she thrives on having lots of friends and fun.

HINT NO. 2
Now notice what happens after *capital* punishment:

> Cindi is a carefree and somewhat irresponsible girl. She thrives on having lots of friends and fun.

ANSWER TO RIDDLE: a COMMA-SPLICE sentence.

A RUN-ON is the incorrect joining of two or more complete sentences. When a comma is used to join the sentences, the run-on becomes a COMMA SPLICE.

The run-on and the comma splice are probably the most noticeable of all commonly made sentence construction errors.

To avoid run-ons and comma splices, just remember this simple rule:

> One way of correcting run-ons and comma splices is to separate the independent clauses with a period and a capital letter.

Cindi is a carefree and somewhat irresponsible girl. She thrives on having lots of friends and fun.

Study these samples of run-ons and comma splices. Observe how they are corrected by separating the two complete sentences with a period and a capital letter. Three other ways to correct run-ons and comma splices are shown in Units 12 and 13.

RUN-ON AND COMMA SPLICE

Business is all round you it is your favorite hamburger place or bowling alley. (Run-on)
Business is all around you, it is your favorite hamburger place or bowling alley. (Comma Splice)

CORRECT SEPARATION

Business is all around you. It is your favorite hamburger place or bowling alley.

RUN-ON AND COMMA SPLICE

Business is the factory you see from the freeway it is the bank downtown. (Run-on)
Business is the factory you see from the freeway, it is the bank downtown. (Comma splice)

CORRECT SEPARATION

Business is the factory you see from the freeway. It is the bank downtown.

RUN-ON AND COMMA SPLICE

Business is part of our society there is no escape. (Run-on)
Business is part of our society, there is no escape. (Comma splice)

CORRECT SEPARATION

Business is part of our society. There is no escape.

RUN-ON AND COMMA SPLICE

American Telephone and Telegraph Company has equipment worth billions of dollars it is one of the giants of American business. (Run-on)
American Telephone and Telegraph Company has equipment worth billions of dollars, it is one of the giants of American business. (Comma splice)

CORRECT SEPARATION

American Telephone and Telegraph Company has equipment worth billions of dollars. It is one of the giants of American business.

REPLAY UNIT 11

Determine your ability to spot and correct run-ons and comma splices. Write C in the blank for a correct sentence. Otherwise, insert a period and a capital letter to make the correction. Be sure you do not place a period after a dependent group of words. If you do, you will have created a fragment.

Samples Sticks and stones will break my bones/ words will never harm me. ______

Sticks and stones will break my bones. Words will never harm me. ___C___

Although sticks and stones will break my bones, words will never harm me. C

1. The new equipment is being shipped to you at once. You should receive it by the end of the week.

2. Since the new equipment is being shipped to you at once, you should receive it by the end of the week.

3. These statistics deal only with symptoms, they do not reveal the fundamental economic problems.

4. Geriatrics, the study of dealing with problems of the aged, is a relatively new science.

5. Geriatrics is the study of dealing with problems of the aged, it is a relatively new science.

6. The tax return is inaccurate he should ask the accountant to prepare a new one.

7. As the tax return is inaccurate, he should ask the accountant to prepare a new one.

8. Many students are attracted to a business career, they like the challenge it offers.

9. Because of the challenge it offers, many students are attracted to a business career.

10. Each year *Fortune* magazine publishes a list of the top five hundred American corporations.

11. Although there have been many small automobile manufacturers in the United States, only General Motors, Ford, Chrysler, and American Motors have survived and grown big.

12. Many of our corporations of today started in colonial times, they became far bigger than anyone expected.

13. Most large corporations have become multinational. In other words, they have gone international. They do business in many foreign countries.

14. A good example of such a corporation is International Business Machines, it has an office in almost every major city in the world.

15. A good example of such a corporation is International Business Machines, which has an office in almost every major city in the world.

16. The in-service training at this company includes instruction in tape perforation and in computerized typesetting.

17. The job requires someone who can operate folding machines, drills, and stitchers.

18. Entry-level positions in the manufacturing technology field often require a knowledge of die making and tool design.

19. The artist painted four portraits of the President's wife, only one could be found

20. Although the artist painted four portraits of the President's wife, only one could be found.

The answers are on page 311.

READ UNIT 12

JOINING

In Read Unit 11 you practiced separating the independent parts (or complete sentences) of run-ons and comma splices. When two independents are closely related, however, joining them is often preferable to separating them. Here are two correct ways to join independent clauses.

JOINING INDEPENDENTS

- a semicolon (;) only.

  Cindi is a carefree and somewhat irresponsible girl; she thrives on having lots of friends and fun.

or

- a comma followed by **and**, **but**, **or**, **nor**, or **for*** (conjunctions).

  Cindi is a carefree and somewhat irresponsible girl, and she thrives on having lots of friends and fun.

Here are the same runs-ons and comma splices as in Unit 11. This time you'll see how they can be corrected by *joining* them *correctly*, instead of *separating* them as in Unit 11.

RUN-ON Business is all around you it is your favorite hamburger place or bowling alley.

COMMA SPLICE Business is all around you, it is your favorite hamburger place or bowling alley.

CORRECT Business is all around you; it is your favorite hamburger place or bowling alley.
Business is all around you, for it is your favorite hamburger place or bowling alley.

RUN-ON Business is the factory you see from the freeway it is the bank downtown.

COMMA SPLICE Business is the factory you see from the freeway, it is the bank downtown.

CORRECT Business is the factory you see from the freeway; it is the bank downtown.
Business is the factory you see from the freeway, and it is the bank downtown.

**For* is a joining word when it means "because."

RUN-ON Business is part of our society there is no escape.

COMMA SPLICE Business is part of our society, there is no escape.

CORRECT Business is part of our society; there is no escape.
Business is part of our society, and there is no escape.

THE TRANSITION TRAP Transitions are words or expressions that skillful writers use to help the reader cross over from one idea to the next closely related idea. Some of the most common **transition words** are:

also	however	moreover	then
for example	in addition	nevertheless	therefore
hence	in fact	otherwise	thus

> When a transition is used between two complete sentences, do use a semicolon; DON'T use a comma *before* the transition word. The result would be a comma splice. DO use a comma between two complete sentences when one of the five proper joining words is also used (**and, but, or, nor, for**).

The RIGHT samples below show how to correct the comma splices by either separating as explained in Unit 11 or correctly joining as shown in this unit. The **transition words** are underlined.

COMMA SPLICE About 50 percent of all those employed in business are engaged in the distribution of goods and services, however, about 40 percent are in production.

Right About 50 percent of all those employed in business are engaged in the distribution of goods and services; however, about 40 percent are in production.

Right About 50 percent of all those employed in business are engaged in the distribution of goods and services. However, about 40 percent are in production.

COMMA SPLICE Please order the new equipment first, then you can inquire about the training program.

Right Please order the new equipment first; then you can inquire about the training program.

Right Please order the new equipment first. Then you can inquire about the training program.

Right Please order the new equipment first, and then you can inquire about the training program.

> When the transition word has more than one syllable, use a comma after it. Do not use a comma after *also* and the one-syllable transitions, such as *thus, then,* and *hence.* Full information about correct punctuation is presented in Sections 9 and 10. For now, just remember to use a semicolon or period directly *before* a transition expression that joins two **independent** word groups.

REPLAY UNIT 12

Spot and correct run-ons and comma splices.

Part A

Write C in the blank for the one correct sentence. For the others, use the semicolon (;) method of correcting the one run-on and the five comma splices. Write *R* in the blank for the run-on and *CS* for the comma splices.

1. First you should discuss the plan with Mrs. Washington then you should give the instructions to Mr. Adams. ____
2. The electronic typewriter is easy to use, however, you will need a few days of practice to become acquainted with it. ____
3. These statistics deal only with symptoms, therefore, they do not reveal the fundamental economic problems. ____
4. Students are extremely fond of homework, for that reason I provide them with many long assignments. ____
5. The tax return is inaccurate, he should ask the accountant to prepare a new one. ____
6. The electronic typewriter is easy to use, but you will need a few days of practice to become acquainted with it. ____
7. Be sure to include all the steps necessary to complete the job, also be sure to list them in the correct order. ____

Part B

Write C for correct, R for run-on, and CS for comma splice. Then correct the run-ons or comma splices by using a comma and one of the five joining words. Choose the word that will make the sentence sound sensible.

8. Mr. Evans was able to repair the motor quickly he is an expert mechanic with many years of experience. ____
9. He didn't notice any of the errors, nor would he have known how to correct them if he had. ____
10. The personnel department is on the third floor the interviewers are out to lunch now. ____
11. The new equipment is being shipped to you at once, you should receive it by the end of the week. ____
12. Sticks and stones will break my bones. Words will never harm me. ____
13. Sticks and stones will break my bones, words will never harm me. ____

14. He noticed the errors, he didn't know how to correct them. ______

15. Either Ms. Jon was the most unhappy woman in the information processing center, or she put on a very good act. ______

Part C

16. List five commonly used transition expressions:

______ ______ ______

______ ______

17. When a transition expression is between independent clauses, you may join the independents in the following way: Insert a comma plus *and, but, or, nor,* or *for* before the transition word (True or False)

18. When a transition expression is between independents, the independents may be separated as follows: Place a period after the first independent and capitalize the first letter of the transition. (True or False)

19. When a transition expression is between independents, the independents may be joined by placing a semicolon before the transition. (True or False)

20. In the following sentence, it would be correct to replace the comma with a semicolon before the word "however." (True or False)

    He will, however, study the blueprint.

21. No. 20 is false because "however" is not used between i______ but merely separates the helping verb from the main verb.

22. Being caught in the Transition Trap results in a r______ or a c______ ______.

Please check your answers on page 311.

READ UNIT 13

LEANING

Often a dependent person will lean on someone who is independent in order to feel fulfilled or complete. When a dependent word group (or clause) can lean on an independent word group, the result is a complete sentence.

Therefore, still another way of correcting a run-on or a comma splice is to change one of the independents into a dependent. A dependent conjunction before an independent word group makes that word group dependent (see Read Unit 9).

Often, however, the dependent conjunction changes the meaning of the original sentence. The choice of whether to use two independent word groups or one independent and one dependent is based on how you want to express a particular idea.

When AND, BUT, OR, NOR, or FOR is the conjunction joining two independents, both parts of the sentence remain independent and of equal importance. However, when a dependent conjunction is used, one of the word groups becomes dependent and appears less important. Typical dependent conjunctions follow:

after, although, as, as if, as though, because, before, even though, if, in order that, provided, since, so that, that, though, unless, until, when, where, whether, which, while (see Read Units 5 and 9)

RUN-ON The electronic typewriter is easy to use you will need a few days of practice to become acquainted with it.

COMMA SPLICE The electronic typewriter is easy to use, you will need a few days of practice to become acquainted with it.

CORRECT DEPENDENT LEANS ON AN INDEPENDENT
Although the electronic typewriter is easy to use, you will need a few days of practice to become acquainted with it.

RUN-ON The tax return is inaccurate Ms. Keogh should ask the accountant to prepare a new one.

COMMA SPLICE The tax return is inaccurate, Ms. Keogh should ask the accountant to prepare a new one.

CORRECT DEPENDENT LEANS ON INDEPENDENT
Since the tax return is inaccurate, Ms. Keogh should ask the accountant to prepare a new one.

RUN-ON First you should discuss the plan with Mrs. Washington then you should give the instructions to Mr. Adams.

COMMA SPLICE First you should discuss the plan with Mrs. Washington, then you should give the instructions to Mr. Adams.

CORRECT DEPENDENT LEANS ON INDEPENDENT
You should discuss the plan with Mrs. Washington **before** you give the instructions to Mr. Adams.

RUN-ON These statistics do not reveal the fundamental economic problems they deal only with symptoms.

COMMA SPLICE These statistics do not reveal the fundamental economic problems, they deal only with symptoms.

CORRECT DEPENDENT LEANS ON INDEPENDENT
These statistics do not reveal the fundamental economic problems **because** they deal only with symptoms.

When joining the parts of a sentence by making one of the clauses dependent, decide which part to make dependent based on how you want to express the idea. In the CORRECT sentences about the electronic typewriter and the tax return, the first part was made dependent by starting the sentence with a dependent conjunction. Notice the comma between the sentence parts:

> **If** the tax return is inaccurate, Ms. Keogh should ask the accountant to prepare a new one.
>
> **Although** the electronic typewriter is easy to use, you will need a few days of practice to become acquainted with it.

However, in the CORRECT sentences about the plan and the statistics, the dependent part comes *after* the independent. The joining is accomplished by inserting a dependent conjunction between the two word groups. Usually a comma is not used before a dependent conjunction. (Correct comma placement is explained in Section 9.)

> These statistics do not reveal the fundamental economic problems **because** they deal only with symptoms.
>
> You should discuss the plan with Mrs. Washington **before** you give the instructions to Mr. Adams.

HOW WILL YOU KNOW WHETHER TO JOIN OR TO SEPARATE?

Two closely related ideas are, in a way, like people who have to decide whether to be joined or separated. No outsider can make the decision for them, although the two people might receive guidance from others.

When writing a business communication, you must be the one to decide whether to join correctly or to separate two closely related ideas. Writing practice, reading well-written business communications, and your own good judgment will enable you to make these decisions quickly.

A technique of skillful business writers is to use dependent clauses for deemphasis of ideas. Applying the same principle, independent clauses are used to *emphasize* information or ideas. That is, use the independent clause for the point you want to give extra importance.

REPLAY UNIT 13

Part A

Write *R* for run-on, *CS* for comma splice, *C* for correct sentence. Underline the six dependent conjunctions.

______ 1. Ms. Hicks buys a new car every year, she saves money by doing that.

______ 2. Because Ms. Hicks buys a new car every year, she saves money.

______ 3. You have broken one of our important rules the warden has promised to give you another chance.

______ 4. You have broken one of our important rules, but the warden has promised to give you another chance.

______ 5. Even though you have broken one of our important rules, the warden has promised to give you another chance.

______ 6. Minnesota must be a big state it raises so much grain.

______ 7. Minnesota must be a big state because it raises so much grain.

______ 8. Minnesota must be a big state, for it raises so much grain.

______ 9. Our peach trees have grown beautifully you ought to see them.

______ 10. Our peach trees have grown beautifully, you ought to see them.

______ 11. Although our winters are long and cold, our summers are short and sometimes very hot.

______ 12. Our winters are long and cold, but our summers are short and sometimes very hot.

______ 13. Since he was a native of the place, Thoreau knew the neighborhood of Concord thoroughly.

______ 14. Thoreau knew the neighborhood of Concord thoroughly he was a native of the place.

______ 15. Although a rolling stone gathers no moss, it does get a certain smoothness from its rolling.

Part B

Change these comma splices into correct sentences by making one of the word groups dependent. If necessary, leave out a word or two. Use one of the following conjunctions in each sentence:

when, because,
even though, although

16. I do not believe in shouting myself hoarse, the team is playing its best.
17. He places big orders with us, that's why the manager is offering him a special discount.
18. The plant was operating on a 24-hour basis, top management refused to adopt a three-shift schedule.
19. The professor explained that studying English is fun, the new student wouldn't believe it.
20. The method of shipment was not vital, however, we gave it careful consideration.

Check your answers on page 311.
Now take the POP QUIZ on page 266.

READ UNIT 14

THE MISSING KEY

As you read the letter on the opposite page and the other samples of well-written business letters throughout this book, observe the simple and easy-to-read style used by good business writers.

BUT . . . THERE ARE SOME PROBLEMS WITH THIS LETTER. The typewriter used for the letter had one key missing and a shift key that didn't always work when it should. That's why the typist used commas not only where they belong, but also wherever the missing key should have been used. Use your pen to repair the damage without changing or adding any words.

Then study the appearance of the letter. Notice the various parts of the letter and their placement on the page:

letterhead
date
inside address
salutation
body
complimentary close
writer's name
title
typist's initials

Numbers to the left of the body will enable you to check your corrections with Replay answers on page 312.

JOHN WILEY & SONS, INC., PUBLISHERS
605 THIRD AVENUE, NEW YORK, N.Y. 10016 (212) 867-9800
TELEX 12-7063 CABLE: JONWILE

Letterhead

December 9, 198- — Date

Professor Robert Wrenn
Business Division
Los Angeles Harbor College — Inside Address
1111 Figueroa Place
Wilmington, CA 90744

Dear Professor Wrenn: — Salutation

Thank you very much for the time and courtesies you extended to our representative, Bob Robinson, when he was at your school last month, he enjoyed his visit with you,

At Mr, Robinson's request we have sent you the new edition of THE WILEY OFFICE HANDBOOK, this was sent to you several weeks ago, and you should have it by now, we do hope you have an opportunity to look it over carefully,

Your name has also been placed on our mailing list to receive an examination copy of PRINCIPLES OF BUSINESS COMMUNICATION, this book, by Pearce, Figgins, and Golen, is expected to come off the press sometime next month, — Body

Your copy will be sent to you just as soon as it becomes available, if there is any way Wiley can be of service to you, Professor Wrenn, please let us know, best wishes for a happy holiday season,

Complimentary Close — Cordially,

Peggy Brigg

Writer's Name / Title — Peggy Brigg
Regional Vice President

lrs — Initials

PS Don't write run-on sentences, they are hard to read,

Checkpoint

Place a check next to each item below when you are sure you understand it.

1. A clause is a word group containing a subject-verb combination:

   we didn't enjoy the show

2. An independent clause may be used as a sentence:

   We didn't enjoy the show.

3. A semicolon is one correct method of joining two independents:

   We didn't enjoy the show; the acoustics were bad.

4. Another way to join independents correctly is to insert a comma followed by one of the following conjunctions: *and, but, or, nor, for*:

   We didn't enjoy the show, for the acoustics were bad.

5. A dependent conjunction may be used for correct joining of clauses. Sometimes the dependent conjunction precedes the first clause of the sentence, in which case a comma joins the clauses.

   We didn't enjoy the program because the acoustics were bad.

   BUT

   Because the acoustics were bad, we didn't enjoy the program.

6. A fragment is a group of words that expresses an incomplete or dependent idea; it masquerades as a sentence because it begins with a capital letter and ends with a period.

   Because the acoustics were bad. FRAGMENT

7. A run-on is the combination of two or more independent and complete ideas into one sentence without punctuation or a conjunction.

   We didn't enjoy the program the acoustics were bad. RUN-ON

8. A comma splice is the use of a comma to separate the parts of a run-on.

   We didn't enjoy the program, the acoustics were bad. COMMA SPLICE

9. Fragments, run-ons, and comma splices are unacceptable in business writing.

SPECIAL NOTE
An experienced writer with a clear understanding of sentence structure will occasionally use a fragment to create a certain effect, particularly in advertising material. After you are recognized as an expert in business writing, you may occasionally use a fragment. UNTIL THEN, DON'T LET FRAGMENTS CREEP INTO YOUR WRITING.

In conversation, fragments are frequently used and are acceptable. Imagine this conversation:

> *Instructor:* What did you study in Units 7 through 9 of this section?
> *Student:* The difference between sentences and fragments.

If the student were to *write* his reply, he would insert a subject and a verb. This would change the student's *acceptable conversational fragment* to an *acceptable written sentence*. Write the acceptable written reply in the blank below.

Have you used the answer section in the back of the book to verify your answers to Replay Units 7–13? *Definitions of simple, compound, and complex sentences and additional grammar explanations are in the Grammar Supplement on pages 297.*

Proofreading for Careers

When *you* are typing, proofread each page before removing it from the typewriter. It is insufficient to rely on "sensing" when you have made a typographical error. Additional errors can be made—errors that are not immediately "felt" while typing.

When someone is typing *for* you, submit, sign, or OK the document only after you have proofread it. Don't rely on anyone else's accuracy or skill when a document is your responsibility.

Everyone makes errors, but competent people find and correct their own before anyone else finds them.

Make the necessary corrections in this short article. Use your dictionary.

CORPORATE ORGANIZATION: A VIEW FORM THE TOP

The owners of a corporation are called stockholders or shareholders. Each stock holder have one vote for each share of stock he or she owns. Every year at a special meeting, the stock holders elect a board of directors. The board represents the stockholders and elects the high-ranking corperate officers, or at least the president. The officers are responsible for the day-by-day operation the of corporation, boreds of directors usually have between 7 and 17 members.

The board of directors, the officers, and the stockholders are require by law to meet at least once eacy year. The corporate secretary notifies the stockholders of the meeting by male. It is common practice to enclosre a proxy form with the notice of the meeting, use of such a form allows the stockholder to submit his vote in lieu of attending the meeting in person.

TOTAL ERRORS=13. Your score is ___.
Circle one: Excellent, Good, Fair, Ugh

Special Assignment

1. Do the Practice Quiz on the following page. Verify the answers.
2. Select the fragments from the Practice Quiz. In the blank at the left, write the number of the fragment. Then write a complete sentence to correct the fragment.

Example 3 The law authorizing this activity was passed in 1975. (1)

______ ____ (2)

____

______ ____ (3)

____

______ ____ (4)

____

______ ____ (5)

____

______ ____ (6)

____

______ ____ (7)

____

______ ____ (8)

____

______ ____ (9)

____

______ ____ (10)

3. Now select the run-ons and comma splices from the Practice Quiz. In the blank at the left, insert the number of the sentence. Then correct it by either separating or correctly joining the independent parts.

Sample

2 Mr. Johnson is the new accountant in our office, but he is not a CPA. (1)

_____ ____________________ (2)

_____ ____________________ (3)

_____ ____________________ (4)

_____ ____________________ (5)

_____ ____________________ (6)

_____ ____________________ (7)

_____ ____________________ (8)

Submit this assignment to your instructor on ______________. (date)

Practice Quiz

Write *A* in the blank next to each correct sentence, *B* next to each fragment, *C* next to each run-on, and *D* next to each comma splice.

_____ 1. The need for an improved worldwide food reserve policy. (Unit 8)

_____ 2. Mr. Johnson is the new accountant in our office he is not a CPA. (Unit 11)

_____ 3. The law authorizing this activity. (Unit 8)

_____ 4. Although Mr. Johnson is an accountant, he is not a CPA. (Unit 13)

_____ 5. To show the exact number of hours before school is out. (Unit 9)

_____ 6. The carpenter is working very hard, nevertheless we don't think he can finish by one. (Unit 12)

_____ 7. While the carpenter is working hard. (Unit 9)

______ 8. While the carpenter is working hard, we don't think he can finish the job on time. (Unit 13)

______ 9. Although his writing is better than mine. (Unit 9)

______ 10. His writing is better than mine, but I speak better than he does. (Unit 12)

______ 11. His writing is better than mine however, I speak better than he does. (Unit 12)

______ 12. Mr. Burns knows her he also knows her sister. (Unit 11)

______ 13. Mr. Burns knows her he met her last week. (Unit 11)

______ 14. If Mr. Burns knows her (Unit 9)

______ 15. Since Mr. Burns knows her well, he might invite her to join the club. (Unit 9)

______ 16. That capable and intelligent young woman from your office. (Unit 10)

______ 17. That young woman from your office is both capable and intelligent.

______ 18. Is going to take a trip to Europe. (Unit 8)

______ 19. She flew to Spain last year, then she took a Mediterranean cruise. (Unit 12)

______ 20. Ms. Naomi Simon is a capable and intelligent woman we need more nurses like her. (Unit 11)

______ 21. Mrs. Powell, carrying the coffee in one hand and the keys in the other. (Unit 10)

______ 22. Carolinda and Eric are our most talented writers in fact, they have won several awards. (Unit 12)

______ 23. Do you want to go?

______ 24. Even though you want to go (Unit 9)

______ 25. About an hour later the students arrived.

SECTION 3

AIN'T IS IN THE DICTIONARY

After Completing This Section, You Will:

- *Locate the various kinds of information included in your dictionary.*
- *Interpret the symbols, abbreviations, and special terms used in your dictionary.*
- *Choose English usage based on information provided in your dictionary.*
- *Spell more accurately than you did before.*
- *Refer to your dictionary frequently as a valuable resource.*

How do words get into the dictionary to start with? Who makes the decisions? When Moses received the Ten Commandments, did he also receive a list of spelling rules? Then why can't I spell "enough" with just four letters—*enuf*?

Scholars called lexicographers review tremendous numbers of words spoken and written throughout the country. These dictionary makers, or lexicographers, report how people use the language. They listen to and observe how all kinds of people write and speak, both formally and informally. They also observe language patterns in various parts of the country and in different types of work—from armed robbery to zoology.

The lexicographers are trained to recognize newly coined words, old words used in new ways, new spellings or pronunciations of words, and even slight changes of meaning for old words. Each word of interest is recorded, along with information on how the word was used, by whom, and the date. Several thousand entries may be added to the file or programmed into a computer daily. These files are so valuable that they may be insured for over a million dollars. When it is time to print the new edition of a dictionary, the entries are sorted to determine which new words should be included and which old words should be changed.

Thus your dictionary provides a *description* of how Americans actually use English. As we all know, nonstandard expressions such as "ain't" are used. Therefore, contrary to the slogan you may have heard in your childhood ("*Ain't* ain't in the dictionary"), "ain't" *is* in the modern dictionary that *describes* rather than *prescribes.*

This modern dictionary doesn't prescribe—or inform you of exactly what to do—as your doctor does when giving you a prescription. You, not your dictionary, make the decision about how to spell, pronounce, or use a word.

Then what good does it do to look up a word in the dictionary? When you interpret the various symbols and abbreviations, you can use your dictionary to make intelligent choices. For example, when you look up the pronunciation of "hospitable," you'll find that there are two pronunciations. You decide which to use, keeping in mind that the first one is usually preferred because it is used more frequently by well-educated people.

Knowing the information given in the dictionary about "ain't" ensures that a job applicant would not use it when applying for a job. However, this same person might write, "He ain't here," as part of the dialogue in a story submitted to the English teacher.

TYPES OF DICTIONARIES

Various types of dictionaries are available: medical, shorthand, slang, synonyms and antonyms, foreign language, children's, and so on.

In addition, spelling dictionaries, which list spelling and syllables only, provide quick reference for the most commonly used words. If you keep one of these mini-dictionaries on your desk alongside your comprehensive desk dictionary, you'll save time when checking spelling or syllables. Here are some titles:

Word Division Manual, by Silverthorn and Perry
20,000 Words, by Leslie
The Word Book, by Ellis
The Word Book, by Kahn and Mulkerne

A thesaurus (pronounced **the SOR us*) is another useful reference book for people interested in words. Have you ever been writing something and just couldn't think of the right word to express what you wanted to say? If you keep a thesaurus on your desk, then you immediately reach for it at such times.

I suggest that if you don't own one, you look at a few the next time you're in the library or a bookstore. They're available in hard cover and paperback. Here's a sample entry from *Webster's Collegiate Thesaurus*:

glad *adj* **1** characterized by or expressing the mood of one who is pleased or delighted <he was *glad* to be on vacation>
syn happy, joyful, joyous, lighthearted
rel delighted, gratified, pleased, rejoiced, tickled; blithe, exhilarated, jocund, jolly, jovial, merry; gleeful, hilarious, mirthful
idiom filled with (*or* full of) delight
con blue, dejected, depressed, downcast, melancholy; despondent, dispirited, heavyhearted, sadhearted, unhappy; forlorn, joyless, sorrowful, woeful
ant sad
2 full of brightness and cheerfulness <a *glad* spring morning>
syn bright, cheerful, cheery, radiant
rel beaming, sparkling; beautiful; genial, pleasant
con dark, dim, dull, gloomy, somber

Section Three deals with use of the comprehensive English language dictionary. Since we live in a rapidly changing society, our language is alive and changing rapidly also. To determine how up to date your dictionary is, check the copyright date.

Here are the names of some good, modern comprehensive dictionaries that are suitable for personal use:

*In this word *th* is pronounced as in *th*ink, not as in *th*is.

American Heritage Dictionary of the English Language, College Edition
Funk & Wagnalls Standard College Dictionary
Random House College Dictionary
Webster's Ninth New Collegiate Dictionary (or latest)
Webster's New World Dictionary of the American Language, College Edition
World Book Dictionary (two volumes)

READ UNIT 15

GETTING TO KNOW YOU

Do you remember, the song "Getting to Know You" from the play and movie *The King and I*? Because the dictionary is the single most valuable reference tool, think of it as a friend whom you wish to get to know better.

Comprehensive dictionaries are available in three general sizes, each of which is recommended for a specific purpose:

ABRIDGED: POCKET DICTIONARY (CONDENSED)

This is usually a paperback with around 50,000 entry words, along with somewhat limited explanations of each word. A small amount of introductory material and possibly a brief appendix are included. Because it is so convenient to carry to classes, a student should own a pocket dictionary.

ABRIDGED: DESK OR COLLEGE DICTIONARY (CONDENSED)

This is usually a hard-cover book containing about 150,000 entry words with fairly complete explanations about each word. Lengthy front matter (introductory materials) and appendixes (back matter) are parts of a good college dictionary. Everyone interested in using English effectively should have one at home and at work.

UNABRIDGED DICTIONARY (NOT CONDENSED)

This may contain between 350,000 and 450,000 entry words—or all the words the lexicographers feel belong in a comprehensive dictionary. The explanatory material for each entry is extensive, as are the front matter and appendixes. Unabridged dictionaries are found in libraries, schools, newspaper offices, and similar places; there is no need to own one unless you are a professional writer or linguist.

No dictionary is a complete listing of all the words in the language. *Webster's Third New International Dictionary* (the most prestigious American dictionary) has about 450,000 entries. However, several million insects, at least a million different kinds of animals, and about half a million varieties of plants have been named. Dictionaries for general use simply have no room for all these words.

In order to use your dictionary effectively and receive the maximum benefit from it, you should know exactly what different kinds of information and aids are included.

FRONT MATTER AND APPENDIXES

Front matter is everything before the "A" listing, and appendixes follow the "Z" listings. Front matter includes details about how to interpret the information given about each entry. A variety of other information is presented in both the front matter and in the appendix.

There may be sections dealing with the history of English; grammar, spelling, and punctuation rules; notes about language usage; dialects; colleges and universities; geographic names with brief descriptions; foreign expressions; abbreviations; biographical names; first names and their meanings; and so on.

ENTRIES

These are the alphabetically arranged words (or word parts, such as prefixes and suffixes), together with the information supplied for each:

- Every entry includes spelling, syllables, pronunciation, definitions, parts of speech.

pronunciation — part of speech

spelling and syllables — **cap·i·tal·ism** \ˈkap-ət-ᵊl-ˌiz-əm, ˈkap-tᵊl-, *Brit also* kə-ˈpit-ᵊl-\ *n* (1854) : an economic system characterized by private or corporate ownership of capital goods, by investments that are determined by private decision rather than by state control, and by prices, production, and the distribution of goods that are determined mainly by competition in a free market — definition

- Some entries have one or more of these: etymology (history of the word), label (in what context the word is usually used), illustrations, synonyms, derivatives (for example, "accidentally" might be included within the entry for "accidental"), plurals, capitalization, and other helpful notes.

brain (brān) *n.* **1. a.** The portion of the central nervous system in the vertebrate cranium that is responsible for the interpretation of sensory impulses, the coordination and control of bodily activities, and the exercise of emotion and thought. **b.** A functionally similar portion of the invertebrate nervous system. **2. brains.** Intellectual capacity. **3.** *Informal.* A highly intelligent person. **4.** Often **brains.** The supreme planner, as of a movement. **5.** An automatic device, as a computer, that is central to a computation or control process. —*tr.v.* **brained, brain·ing, brains. 1.** To smash in the skull of. **2.** *Slang.* To hit on the head. —***idioms.* on the brain.** Obsessively in mind. **rack (**or **beat) (one's) brains.** To think as hard as one can. [ME < OE *brægen.*]

usage label — etymology

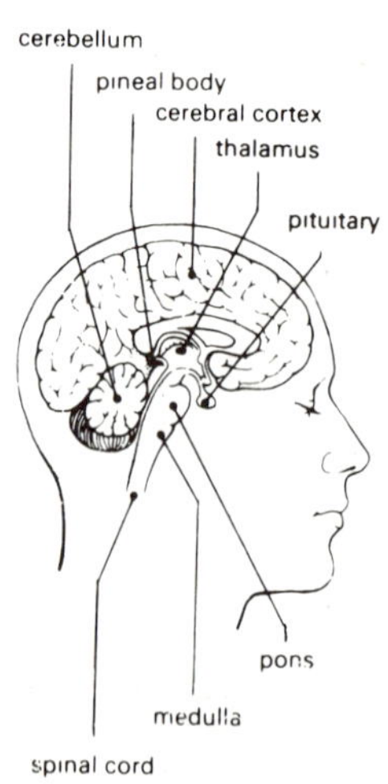

brain

AIDS FOR EFFICIENT USE OF DICTIONARY

- **Guide words** at the top of each page enable you to locate words quickly because they show the alphabetic range of that page. The guide word on the left is the first entry word of that page, and the guide on the right is the last entry.
- The **pronunciation key** is at the bottom of every page or on alternate pages. A complete pronunciation key is also found in the front matter.

REPLAY UNIT 15

1. What is the name of your desk dictionary? ______

   What is the copyright date (most recent)? ______

2. What is the name of your pocket dictionary? ______
3. What are the words called that are at the top of each page?

   GUIDE WORDS

4. Open your dictionary to any page. Then look at the page on your left and write the guide words in these blanks.

   ______ ______

5. What is the purpose of the guide words? to show the alphabetic range of the page
6. What information can you find out about every entry word?

   Spelling | Syllables | Pronunciation
   definitions | Parts of a speech

7. What other types of information are included with some entry words?

   etymology | usage | illustration | synonyms
   derivations | plurals | Capitalization

8. A comprehensive dictionary containing at least 350,000 entries is known as a/an Unabridge Dictionary dictionary.
9. The other two types of comprehensive dictionaries are known as Desk or college and Pocket.
10. If you are interested in using English effectively, you should have a Desk or college dictionary at home and at work.
11. If, in addition, you are a student or for some other reason need to carry a dictionary with you often, you should have a Pocket dictionary.
12. What do you usually find at the bottom of every page or every other page of a good dictionary? Pronunciation key

13. Open your dictionary to a randomly selected page, see if you find what you named in No. 12, and check one of these blanks:

    Yes, it's there, ________________ No, it isn't, ________________

14. Details about how to interpret the information given about each entry word can be found in the Front matter.

15. Look at the appendix items (or back matter) listed in the table of contents of your desk dictionary. Then thumb through the appendix. What are the main titles of each section in the appendix of your desk dictionary? ________________

    ____

    ____

The answers to this Replay are on page 312.

READ UNIT 16

CRACKING THE CODE

To save space, a code—made up of symbols, special words, and abbreviations—is used in dictionaries. Cracking this code is essential if you are to use the dictionary intelligently. As an educated person, you should know how to find and interpret any information provided in your dictionary. In fact, one of the most important characteristics of a well-educated person is that he or she knows how and where to look up information.

The code varies somewhat from one dictionary to another. Here are some explanations of the codes given in the introductory pages of the *American Heritage Dictionary* and *Webster's Ninth New Collegiate Dictionary*. The abbreviations AHD and WNCD refer to these two dictionaries. For details that may differ in YOUR dictionary, see the front matter.

SPELLING AND VARIATIONS

(a) The entry word is the correct spelling of the word. (The dots in the entry word indicate syllables.) If there is a second correct spelling, it will follow a comma or the word "or":

(Etymology in brackets **precedes** definitions in WNCD.)

[1]**the·ater** *or* **the·atre** \'thē-ət-ər, 'thi-ət-, *oftenest in Southern* 'thē-ˌāt- *also* thē-'āt-\ *n* [ME *theatre,* fr. MF, fr. L *theatrum,* fr. Gk *theatron,* fr. *theasthai* to view, fr. *thea* act of seeing; akin to Gk *thauma* miracle] (bef. 12c) **1** **a :** an outdoor structure for dramatic performances or spectacles in ancient Greece and Rome **b :** a building for dramatic performances **c :** a building or area for showing motion pictures **2 :** a place of enactment of significant events or action ⟨the ~ of public life⟩ **3** **a :** a place rising by steps or gradations ⟨a woody ~ of stateliest view —John Milton⟩ **b :** a room often with rising tiers of seats for assemblies (as for lectures or surgical demonstrations) **4** **a :** dramatic literature : PLAYS **b :** dramatic representation as an art or profession : DRAMA **5 :** dramatic or theatrical quality or effectiveness

(Etymology in brackets **follows** definitions in AHD)

ax or **axe** (ăks) *n., pl.* **ax·es** (ăk′sĭz). **1.** A tool with a bladed head mounted on a handle, used for felling or splitting lumber. **2.** A tool or weapon, such as a battle-ax, similar to an ax. **3.** *Informal.* **a.** A sudden termination of employment: *got the ax yesterday.* **b.** An abrupt or ruthless removal. —*tr.v.* **axed, ax·ing, ax·es. 1.** To use an ax. **2.** To remove ruthlessly or suddenly. —***idiom.* have an ax to grind.** To pursue a selfish or subjective aim. [ME < OE *æxa.*]

Although either spelling is correct, choose the first for business and professional writing, because the more frequently used form is usually first.

(b) When the word "also" separates the two spellings, the second one is less acceptable:

lov•able *also* **love•able**

(c) Some words are spelled differently in British English. This variation is shown by means of a "label," such as "chiefly British."

col•or *Also chiefly British* **col•our**

(d) Sometimes the second spelling is listed as a separate entry word. You'll know this is not the "preferred" form because in place of where the definition should be, you'll find the preferred spelling.

al·right (ôl-rīt′) *adv. Nonstandard.* All right.

al·right \(ˈ)ȯl-ˈrīt, ˈȯl-ˌ\ *adv or adj* [ME, fr. OE *ealriht*] (14c) : ALL RIGHT
usage In now obsolete senses *all right* or *alright* was formed in Old English as *ealriht.* Variation in early scribal and printing practices and in spoken stress patterns has given us this and similar pairs in *all ready, already* and *all together, altogether.* Since the 19th century some have insisted that *alright* is wrong, but, though it is less frequent than *all right,* it remains in common use and appears in the work of reputable writers ⟨the first two years of medical school were *alright* —Gertrude Stein⟩ ⟨it is doing a bit of *alright* —P. H. Dougherty (*N.Y. Times*)⟩

You would then refer to the entry spelled "all right," where you would find the definition.

(e) Compound expressions are entries consisting of two or more words combined to express one concept:

real estate
handmade
hand-me-down

Please look up REAL ESTATE and notice the space between REAL and ESTATE. Now look up HANDMADE. The dot or accent mark between the syllables is how you know it is spelled as one solid word. When you look up HAND-ME-DOWN, you see hyphens between the parts, making it clear that this word is spelled with hyphens, just as shown in your dictionary.

PRONUNCIATION The pronunciation of a word follows the entry word and is often found in italics (*slanted letters*) or enclosed in parentheses ( ) or diagonal lines \ \.

(a) Special symbols called **diacritical marks** show how the letters are pronounced. These symbols are interpreted at the bottom of each page or every other page in your dictionary. Sometimes more than one pronunciation is given. This may mean that the first is preferred or that they are equally "respectable." Please open your dictionary now and use the **pronunciation key** at the bottom to interpret the symbols, or diacritical marks, (such as - ¨ ˘) for pronouncing APRICOT.

apri·cot \'ap-rə-ˌkät, 'ā-prə-\ *n, often attrib* [alter. of earlier *abrecock,* deriv. of Ar *al-birqūq* the apricot] (1551) **1 a :** the oval orange-colored fruit of a temperate-zone tree (*Prunus armeniaca*) resembling the related peach and plum in flavor **b :** a tree that bears apricots **2 :** a variable color averaging a moderate orange

a·pri·cot (ăp′rĭ-kŏt′, ā′prĭ-) *n.* **1.** A tree, *Prunus armeniaca,* native to western Asia and Africa, widely cultivated for its edible fruit. **2.** The juicy, yellow-orange peachlike fruit of the apricot. **3.** A moderate, light, or strong orange to strong orange yellow. [Alteration of earlier *abrecock* < Port. *albricoque* < Ar. *al-birqūq,* the apricot : *al,* the + LGk. *praikokion* < Lat. *praecoqquus,* ripe early (*prae-,* before + *coquere,* to ripen).]

(b) The **accent marks** also require interpretation. The relative degree of loudness with which the syllables of a word are spoken is indicated by accent marks (which look like apostrophes). There are three degrees of loudness (or stress):

Weak—no accent mark.
Strong—primary accent mark.
Medium—secondary accent mark.

When a word has two or more syllables, the word will have one *primary* accent (or stress) mark. One or more *secondary* accent marks may also appear in words of two or more syllables. Please look up ASTRONAUT in your dictionary.

AHD—The darker accent mark indicates primary stress, and the lighter one, secondary. Both appear AFTER the syllable to be stressed.

as·tro·naut (ăs′trə-nôt′) *n.* A person trained to pilot, navigate, or otherwise participate in the flight of a spacecraft. [ASTRO- + Gk. *nautēs,* sailor < *naus,* ship.]

WNCD—The raised accent mark shows primary stress, and the lowered one, secondary. They both appear BEFORE the syllable to be stressed.

as·tro·naut \'as-trə-ˌnȯt, -ˌnät\ *n* [*astr-* + *-naut* (as in *aeronaut*)] (1929) : a person who travels beyond the earth's atmosphere; *also* : a trainee for spaceflight

If you are using some other dictionary, interpret the code for showing stress. You'll find the explanation in the front matter.

DEFINITIONS

(a) Both WNCD and *Webster's New World Dictionary of the American Language* list the earliest definition of a word first and continue in more or less chronological order. AHD and others give the central meaning of the word first. Look up the word QUIZ.

> AHD—the first definition for the verb is "1. To question closely . . ." Continuing on through the definitions, you find "3. *Chiefly British*. To poke fun at; mock."

> WNCD—The first definition for the verb is "1: to make fun of: mock." The third definition is "3: to question closely."

The reversed order of definitions shown above illustrates the importance of knowing the order of definitions in your dictionary.

(b) Another distinction related to definitions is that in WNCD the etymology (history or derivation of the word) is in dark square brackets BEFORE the definition; in the AHD the etymology appears AFTER the definition, but also within dark square brackets [ ]. Look up SANDWICH in your dictionary and notice where the etymology is.

When looking up a meaning, be alert to the etymology so that you won't carelessly mistake it for a definition. The etymology may tell us the languages a word was derived from, such as Middle English, French, Latin, and so on; or it may tell us how the word originated or came into use—as in the case of the word "sandwich." When the languages of origin are given, they are abbreviated; for example, OE (Old English), ME (Middle English), L (Latin), and F or Fr (French). The abbreviation "f." may be used to mean "from" or "derived from," or the symbol > may be used to mean the same thing.

REPLAY UNIT 16

Please look up the word LEOTARD in your dictionary and answer these questions:

1. How many syllables does it have? ___3___
2. Which syllable has the primary stress? ___1rst___ (1st, 2nd, or 3rd)
3. Which syllable has the secondary stress? ___3rd___ (1st, 2nd, or 3rd)
4. Which syllable has no stress? ___2nd___ (1st, 2nd, or 3rd)

5. What is the etymology of this word? ____________________________

____

6. Using the pronunciation key at the bottom of the page, say the word aloud.

Now look up the word WRECKER and answer these questions:

7. According to the pronunciation key at the bottom of the page in your dictionary, the first *e* is pronounced the same as the *e* in what word?

____

8. In the expression "towed in by the wrecker," what is the definition of WRECKER? ____________________________

____

9. What is the correct spelling of TICKERTAPE? (a) tickertape (b) ticker-tape (c) ticker tape (d) any of these (e) none of these

10. What is the preferred spelling of WOOLY? ____________________________

11. What is (are) the correct spelling(s) of CATALOG? ________________

12. What is the correct spelling of ANTITRUST? (a) antitrust (b) anti trust (c) anti-trust (d) any of these (e) none of these

13. Which syllable should have the primary stress in AFFLUENT? ________

14. How do you say ENTREPRENEUR and what does it mean? ________

____

15. Look up the pronunciation of the following words: amass, polite, placement, bogus, visible. By what pronunciation symbol is each of the italicized vowels represented? __ What words in the pronunciation key illustrate, the sound of that symbol? ____________________

______________ ______________ ______________

Please check your answers on page 312.

NICE TO KNOW
The symbol ə (called a schwa) usually appears in the dictionary pronunciation of the vowel in an unaccented syllable. Regardless of whether the vowel is actually *a, e, i, o,* or *u,* the schwa means that the vowel is pronounced something like "uh."

Can you pick out the schwa sounds in these words?

banana collect easily gallop circus

READ UNIT 17

THE TRANSLATION

In Unit 16 you decoded certain kinds of information in your dictionary. Continue to use your dictionary as you study this unit. Look up each entry word that is given as an example, and translate the symbols and abbreviations into information that is meaningful to you.

SYLLABLES

To determine the syllables, look at the centered dots or hyphens that appear in the entry word. Some dictionaries use accent marks instead of dots.

Syllable marks show where printers should place a hyphen in order to continue the word on the next line. However, word-division custom for typing business communications is based not only on the entry-word syllables but also on the rules in Unit 63 and the "Word Division" section on page 243. Sometimes the syllables shown in the pronunciation differ from the syllables appearing in the entry word, but the entry word is the basic guide for word division.

Please look up MULTIPLICATION and observe the system for showing syllables in your dictionary. Now look up FULL-FASHIONED; notice that the hyphen divides the first and second syllables, and the centered dot divides the second and third. While you're in the *F* section, turn to FURIOUS; the entry word syllables differ from the pronunciation syllables in both AHD and WNCD.

PARTS OF SPEECH

Abbreviations show the usual parts of speech for a word. The names of the parts of speech studied in Section One are used, except that verbs are shown in two categories—transitive and intransitive (explained in Grammar Supplement for Section 6).

When a word is used as more than one part of speech, separate definitions are given for each. Study the entry for BROWN in your dictionary. Look it up right now and notice the abbreviations for the parts of speech that are used in your dictionary. You'll observe that "brown" is defined as a noun, an adjective, and a verb (both transitive and intransitive) and that other words using "brown" as a root (the central meaning of a word) are shown also.

USAGE OR STYLE LABELS

A label is a notation that informs the reader of a special way that word is used. Words, definitions, and spellings are labeled in a number of ways; and the system varies with the dictionary. Most words, however, do not have labels because they are standard English, and there are no particular usage restrictions.

A word labeled "archaic" (or "arch.") would be used only in special contexts, although it was long ago considered a generally appropriate word. Picture a store named

YE OLDE SWEETE SHOPPE

This name consists of the archaic word "ye" and three archaic spellings.

A word labeled "obsolete" (or "obs.") is no longer used at all, and a word labeled "rare" is rarely used. Other examples are subject labels such as *chemistry, biology, medicine*; regional labels such as *British, Scotch, Southwest, Canadian*; and style labels such as *slang, colloquial, substandard, nonstandard, illiterate, dialect, obscene, offensive, vulgar, poetic, informal.* The labels are usually abbreviated.

In AHD the system is explained in the front matter under the heading "Labels" in the section called "Guide to the Dictionary." In WNCD, explanations about labeling are under the heading "Usage" in the front matter section called "Explanatory Notes." In *Webster's New World Dictionary*, the explanation is in the section called "Usage Labels and Notes" in "Guide to the Dictionary."

OTHER INFORMATION

Read the table of contents and scan the front matter and appendix to see what other categories of information are available in your dictionary. Then find the following:

- **Symbols** used in various sciences and math are in the appendix section called "Signs and Symbols" in WNCD and in *Webster's New World Dictionary*. However, in AHD, symbols are in the regular alphabetic listing under the word "Symbols."
- **Abbreviations** such as *i.e.*, *lb*, and *hr.* are in the regular alphabetic listing of *Webster's New World Dictionary*. However, they are in a special appendix section of AHD and WNCD. Explanations of abbreviations that are part of the dictionary code are in the front matter of all dictionaries.
- **Miscellaneous** information such as Morse Code, money (or currency) equivalents for various countries, proofreaders' marks, Roman numerals, weights and measures, chemical elements, musical notes, and punctuation and grammar rules are other features you might find either within the alphabetic listing or in the appendix.
- **Geographical Names and Biographical Dictionary** are in the back of WNCD and AHD. They are, however, in the regular alphabetic listing of *Webster's New World*. In some dictionaries the geographical names section is called "Gazetteer" or "Pronouncing Gazetteer."
- **Synonyms and Antonyms** are sometimes found after the definitions. Synonyms are words with similar meanings. Sometimes synonyms are inaccurately defined as words that mean the same; no two words have exactly the same meaning. When synonyms are provided, they help in understanding the definitions and in finding a word with the exact shade of meaning desired. Antonyms are words with opposite meanings.
- **Foreign Words and Phrases** used in American communication are in the general alphabetic list. The appendix of WNCD, however, has a supplementary list of foreign expressions not often used in American English.

> Inside the front cover of AHD is a key to locating these special reference materials. In WNCD an index on the last page is a useful guide to finding the various sections. Both dictionaries have a table of contents as well.

REPLAY UNIT 17

Here's an opportunity to reinforce the dictionary habit. Use your dictionary to answer these questions. What is important on this page is not the information but your ability to find it. Even if you already know the answer, look it up anyway.

1. Give a synonym for each of these words:

   talkative ____________________

   evident ____________________

   vacillate ____________________

2. Show the correct pronunciation (using accent marks and diacritical marks) and the parts of speech for these words:

	PRONUNCIATION	PARTS OF SPEECH
subtlety	____________	____________
rationale	____________	____________
franchise	____________	____________

3. The word January is derived from the name of the ancient Roman god of gates and doorways. What was his name? ____________

4. What is the etymology of the word SECRETARY? ____________

   ____

5. What is the population of Rhode Island? ____________

6. What do these proofreaders' marks mean? # ____________

   tr ____________ stet ____________

7. What is the capital of Kentucky? ____________

8. Write the number 137 in Roman numerals. ____________

9. When was the statesman Winston Churchill born? ____________

10. What is the basic unit of money (or currency) in Kenya? ____________

11. What does the abbreviation OPEC stand for? ____________

12. Where are these words divided into syllables according to the dictionary? Draw diagonals (/) between the syllables.

    accurate company correspondence vaccination

13. Illiterate expressions, vulgarities, and slang are not found in the better dictionaries. True or False? ____________

14. What usage label, if any, does your dictionary show for these words?

    dogie ____________ anyways ____________ critter ____________

15. What does "colloquial" mean? ____________

    ____

Check answers on page 313.

DID YOU KNOW THAT

The ancient Greeks considered a certain type of purple stone to be a remedy for drunkenness. They called the stone *amethystos* (a = "not" and *methyein* = "to be drunk"). We call this stone the *amethyst*.

The bluish-green gem *turquoise* literally means "Turkish" in French and has this name because it was first brought from Turkey.

READ UNIT 18

MISSPELLERS ANONYMOUS

If you've ever wished to be a member of Misspellers Anonymous, here's your chance to improve your spelling and remain anonymous; no one has to know about that "spelleton" in your closet.

- Being an adequate speller takes no special talent. MOST IMPORTANT is to want to be a good speller and to believe you can. WHAT YOU CAN BELIEVE, YOU CAN ACHIEVE.
- For overall spelling improvement, consciously look at words you're reading for about five minutes a day. That is, instead of reading quickly for content, spend just five minutes of your reading time looking at the appearance of words in the newspaper or whatever book or magazine you are reading. That's all it takes to store a tremendous number of words in that magnificent personal computer between your ears.
- When you want to retrieve a word from your computer, relax and write what you think of first. If spelling has been a problem for you, you probably remember times when you were taking a spelling test and erased in order to replace the correctly spelled word with an error. In other words, your first hunch is usually right.
- OF COURSE, WHEN YOU'RE IN DOUBT AND CAN USE A DICTIONARY, DO SO. If you often use your dictionary for spelling, it probably means you're a good speller; remember, no one knows your secret. The reader just sees the page with correctly spelled words.
- If a particular word has been troublesome, try to develop a memory device. Here are 15 commonly misspelled words, each with a memory device to correct the part that gives the trouble. Just stare at the word and the memory device; then say the word and the memory device aloud three or four times; you'll know the correct spelling for life.

accommodate: 2 *c*'s, 2 *m*'s; two coat closets in a modern motel.
recommend: re+commend.
privilege: has a leg in it.
bachelor: has an ache because he's sad not to be married.
superintendent: he fixed the dent.
conscience: science is in it.
pursue: pursue the purse snatcher.
pronunciation: a Catholic word because it has a nun.
persistent: your sister has a tent.
congratulate: rat is in the middle.
dilemma: Emma has a dilemma.
embarrassed: Railroad (RR) and steamship (SS).
villain: the villain lives in a villa.
separate: Pa is in the middle.
weird: we are weird.

The sillier the memory device, the more effective it will be.

- If you don't readily think of a memory device for a particular word you want to remember, here's another way to master the spelling of it for life. On a clean sheet of notebook paper, print the word across the page in normal size, but use huge printing for the troublesome letter or letters. Then stare at the page for about 15 seconds. Repeat the staring procedure each day for three days, and you'll never again forget that word.

sepArate

- Keep handy one of those spelling dictionaries I told you about on page 62.
- Although spelling ability is not related to intelligence, most people do tend to view the inaccurate speller as uneducated, careless, or generally incompetent. To increase the number of commonly used business words you can spell confidently, use the 3, 2, 1 Plan explained in "Spelling for Careers" at the back of this book. Your spelling and dictionary usage will also improve as a result of completing the rest of this textbook according to directions given in each section.

A POSITIVE ATTITUDE TOWARD SPELLING AND FREQUENT USE OF A DICTIONARY ARE THE TWO MOST IMPORTANT WAYS TO IMPROVE SPELLING.

What can you do if you can't find a word in the dictionary because you have no idea of how to spell it? Follow these steps.

1. Try several possible letters that may fit in the questionable part of the word; look through two or three pages of your dictionary trying to find it.
2. Look it up in the MISSPELLER'S DICTIONARY published by Random House. If that doesn't work, go to Step 3.
3. Telephone a friend and ask for help. If your friend isn't home, then . . .
4. Think of another word to use instead!

REPLAY UNIT 18

TRUE/FALSE—Write *T* or *F* in the blank.

_______ 1. To be a good speller requires a special talent.

_______ 2. Almost anyone who wants to be a good speller can be a good speller.

_______ 3. Poor spellers are always using the dictionary.

_______ 4. A person can improve spelling while reading.

_______ 5. Most people tend to assume that an inaccurate speller is uneducated and incompetent.

_______ 6. A person who is highly intelligent will just naturally be a good speller.

_______ 7. A list of words commonly used in business appears in the back of this book and may be used for spelling improvement.

_______ 8. A spelling dictionary doesn't include definitions.

_______ 9. One way to memorize the spelling of a word is to print it and exaggerate the size of the problem letter(s).

_______ 10. Another way is to use a silly memory device.

Please fill in the missing letters. Then rewrite the word twice in the blank if it is a problem word for you. Use large printing for your troublesome letters.

11. sep__rate ____________________

12. w____rd ____________________

13. re________end ____________________

14. ac__________date ____________________

15. priv__________ ____________________

16. bach____or ____________________

17. superintend__nt ____________________

18. persist__nt ____________________

19. dil______a ____________________

20. vil______n ____________________

21. cons__________ ____________________

22. p__rsue ____________________

23. congra__ulate ____________________

24. pron__nciation ____________________

25. emba__________ed ____________________

Here are common misspellings for additional words frequently used in business. Write the correct spelling in the blank at least once. Use your dictionary when in doubt.

26. accidently ____________________

27. existance ____________________

28. deductable ____________________

29. fullfilled ____________________

30. heighth ____________________

31. irrevelant ____________________

32. mathamatics ____________________

33. occassion ____________________

34. occured ____________________

35. rediculous ____________________

DID YOU KNOW THAT . . .
CURFEW was derived from an old French word "covrefeu," which meant "cover fire," the fixed hour when the peasants of the Middle Ages were required to extinguish their fires.
SADIST is the word derived from the name of an 18th-century French nobleman, the Count de Sade. He delighted in describing to his friends how cruelly he treated his girlfriends and his employees.

See page 313 for correct answers.
POP QUIZ on Units 15–18 is on page 267.

Checkpoint

What does APROPOS mean and how do you pronounce it? What is a synonym for "indicate"? What's the plural of "shelf"? One of the most often repeated commands in the language may be "Look it up!"

Many people are unaware of the various kinds of information available in a good comprehensive desk or college dictionary. As a well-educated person, be familiar with the contents and organization of the particular dictionary you use. By exploring the front matter and appendixes in your own dictionary, you become aware of what a versatile reference book you have right on your desk; it can often save you a trip to the library. By studying the explanatory notes, you can decipher dictionary code and accurately interpret the information provided in the entries.

Just because a word is heard on television or is used in print in a particular way doesn't mean that it is correct. In fact, just because a word is in a dictionary doesn't mean that the writers of that dictionary approve of the word. Dictionaries describe how people are using the language. Language style books and textbooks such as this provide further information on what is considered good English usage.

English is a lively language that keeps pace with the rapid changes of 20th century life. That's why it's essential to check the copyright date before buying a new dictionary. *Webster's New World Dictionary* added 20,000 new words and meanings and AHD added 25,000 to their latest college editions. Of the 25,000 additions, 10,000 are business-related, office automation, and technical terms.

In the early 1970s many prisoners of war were released from camps in Vietnam and returned to the United States. To help the returning prisoners adjust to changes that took place during their absence, AHD added a glossary to its 1973 edition. Here are a few expressions that were included in this glossary.

> air piracy, area code, busing, consumerism, liberation, Medicaid, Ms., pantsuit, recycle, sexism, tie-dye

These words are now in the regular alphabetic listing of the current edition of the AHD.

Our changing society causes not only the rapid accumulation of new words in our language but also the loss of outmoded words. All this happens so fast that our dictionaries cannot really keep up to date. Although "automobile" is still a respectable word, "car" is now the word in general use. What used to be called a "gear shift" is now a "stick shift." Jack Smith, a columnist for the *Los Angeles Times*, wrote, "They (old words) hang in the dictionary like dead leaves on a tree; nobody bothers to blow them away. In time they simply fall, unheard, unseen, unmourned."

PLACE A CHECK IN THE BLANK WHEN YOU ARE CONFIDENT YOU CAN INTERPRET THE SYMBOLS OR ABBREVIATIONS IN YOUR OWN DICTIONARY FOR THE FOLLOWING:

_______ Spelling and Variations

_______ Pronunciation

_______ Syllables

_______ Parts of Speech

_______ Labels: subject, style, regional

_______ Etymology

_______ **Write a check in the blank** when you know the order in which definitions are listed in your own dictionary.

_______ **Write a check in the blank** after you read the table of contents and scan the front matter and appendixes of your **own** dictionary.

Special Assignment

To the Student of Apothegms:

What, you don't know who, why, or where apothegms are? Neither did I until I looked in my dictionary for a synonym for the common word PROVERB. Short, witty, and instructive sayings are remembered much better than long-winded ones with many-syllabled words. Use your dictionary and your childhood memories to fill in the blanks with the simple words that are remembered. If you get stuck, ask your family and friends to join you in some fun with words. The answer to the first one is shown as a sample.

1. Repetition of an erroneous act will not constitute that which is morally good, equitable, and fitting.

   Two wrongs don't make a right.

2. Precipitance effectuates dissipation.

   ____

3. A plenitude of gourmand-satiating scullions tend to obviate, obstruct, undermine, decompose, and otherwise boggle the mulligatawny.

   ____

4. Meddlesome inquisitiveness on the part of that domesticated carnivorous member of the zoological family *Felidae* has brought about the unfortunate bereaving of its vivifying forces.

   ____

5. A mineral matter of various composition when engaged in periodical revolutions exhibits no tendency to accumulate any of the cryptogamic plants of the class music.

____

6. From deliberative investigation it has come to our attention that the aviatorial member of the phylum Chordata who is anticipatory will invariably apprehend the member of the slender soft-bodied bilateral invertebrates.

____

7. Close acquaintanceship engenders reproachful disdain.

____

8. The inner surface of every visible mass of vapor in the upper atmosphere is covered with a precious metallic chemical element that is very ductile and malleable.

____

9. Using any other appellation to refer to a specimen from the floriferous shrub of the family Rosaceae would not render the perception of it by means of the olfactory nerves to be any less saccharine.

____

10. One of the ramifications of abandoning the celibate condition with precipitateness is to feel regret for what one has done during the period of freedom provided by the cessation of activities.

____

Proofreading for Careers

The letter that follows is a response to a customer's complaint. Please proofread it carefully because it has 19 spelling errors. With the help of your dictionary, correct them. Start by circling the errors.

Dear Ms. Kaye:

Your resent letter was recieved while I was away from the city, and I apologyze for my delay in repling.

I am very sorry to learn of your experence and have asked our Marketing Department to look into the matter. We are grately conserned about the customer relasions of our dealers. Even through they are independant owners who operate stations leased from us, they are, in many instanses, Barko's only contact with the public.

We certinly do not aprove of the manner in which this paticular dealar responded to your complaint. When we have had an opportunity to reveiw all the facts in this situation, we shall be in touch with you.

Thank you for bringing this experence to our attension.

Very truely yours,

Write the correct spellings below.

Paragraph 1

__________ __________ __________ __________

Paragraph 2

__________ __________ __________ __________

__________ __________ __________

Paragraph 3

__________ __________ __________ __________

____

Paragraph 4 and Closing

__________ __________ __________

PROOFREADING SCORE is __ Circle one: Excellent, Good, Fair, Ugh.

Practice Quiz

Use your dictionary to answer these questions. Read all possible answers before selecting the best one.

_______ 1. The verb RECORD should be pronounced with the stress on the (a) 1st syllable (b) 2nd syllable (c) 1st or 2nd syllable (d) secondary accent (e) primary accent. (Units 16 and 17).

_______ 2. Which spelling (or spellings) is preferred? (a) lovable (b) loveable (c) neither lovable nor loveable (d) both lovable and loveable. (Unit 16)

_______ 3. AFTER is what part of speech? (a) preposition (b) adverb (c) conjunction (d) adjective (e) all of these. (Unit 17)

_______ 4. Martin Van Buren died in (a) 1782 (b) 1862 (c) 1837 (d) 1841 (e) the arms of his paramour. (Unit 17)

_______ 5. THEE would be an example of a word that is (a) archaic (b) obsolete (c) British (d) colloquial (e) slang. (Unit 17)

_______ 6. The meteorology symbol ★ means (a) sun (b) star (c) rain (d) snow (e) asterisk. (Unit 17)

_______ 7. The roman numeral XC is the same as the Arabic numeral (a) 100 (b) 110 (c) 60 (d) 90 (e) none of these. (Unit 17)

_______ 8. The abbreviation SEC means (a) a printed word or phrase is not a mistake (b) see East Coast (c) Southeast Corner (d) Securities and Exchange Commission (e) Society for Eagle Conservation.

_______ 9. What does the French phrase "raison d'être" mean? (a) increase in wages (b) the same as the Latin "racemus" (c) raisin pudding (d) reason or justification for existing (e) none of these. (Unit 17)

_______ 10. Guide words appear (a) at the bottom of every page (b) at the bottom of every page or alternate pages (c) at the top of every page (d) only in unabridged dictionaries (e) in the front matter or appendix. (Unit 15)

_______ 11. A short straight horizontal line (¯) over a vowel is an example of (a) secondary stress mark (b) primary accent mark (c) syllable indicator (d) short sounded vowel (e) diacritical mark. (Unit 16)

_______ 12. ETYMOLOGY means (a) part of speech (b) history of a word (c) synonyms and antonyms (d) secondary stress (e) usage label. (Unit 15)

_______ 13. You look up a word in a college dictionary and don't find a usage label. Therefore (a) you're using an inadequate dictionary (b) the word is colloquial (c) you should look it up in an unabridged dictionary (d) the word is standard and there is no special restriction for its use (e) you should refer to the front matter to be sure you're interpreting the code correctly. (Unit 17)

_______ 14. TWINKLING is correctly divided into syllables in which of the following examples? (a) twin•kling (b) twink•ling (c) twinkl•ing (d) twi•nkling (e) twi•nkl•ing. (Unit 17)

_______ 15. SAFETY and DANGER are (a) synonyms (b) homonyms (c) diacriticals (d) colloquials (e) antonyms. (Unit 17)

_______ 16. What is the capital of Sweden? (a) Swedenborg (b) Stocklohm (c) Stockholm (d) Scandinavia (e) Oslo. (Unit 17)

_______ 17. Which word is incorrectly spelled? (a) occured (b) accommodate (c) weird (d) recommend (e) dilemma. (Unit 18)

TRUE/FALSE—Write *T* or *F* in the blank.

_______ 18. A biographical dictionary is found in the appendix of all good dictionaries. (Unit 17)

_______ 19. A dictionary containing close to half a million entries is called an unabridged dictionary. (Unit 15)

_______ 20. Just as the dollar is the name of the unit of exchange in the United States, the pound is the name of the Israeli unit of exchange. (Unit 17)

_______ 21. SUPERINTENDANT is correctly spelled. (Unit 18)

_______ 22. The word in No. 21 has five syllables. (Unit 17)

_______ 23. You would avoid using the adjective WIGHT in a business letter because it is slang. (Unit 17)

_______ 24. The usual sign of a poor speller is frequent use of the dictionary. (Unit 18)

_______ 25. The abbreviation "cwt." means "hundredweight." (Unit 17)

SECTION 4

APPLES, TIGERS, AND SWAHILI

After Completing This Section, You Will:

- *Consistently apply principles of standard English to*
  —forming plurals of regular and irregular nouns.
  —writing compound nouns.
  —capitalizing nouns.
- *Spell correctly and understand the meaning of the words presented in this section.*
- *Avoid using nouns that might be considered "sexist."*

Nouns are the words that name people, animals, places, things, ideas, and actions. When you think about exotic faraway places or your hometown . . . movie stars, kings, and the person next door . . . life, health, peace, success . . . monopoly, jogging, and homework . . . apples, records, or a Rolls Royce—you are thinking NOUNS. Look around and you'll see nouns.

Reading, listening, and general experience enable you to use a great many nouns correctly most of the time. The study of a few simple rules plus intelligent use of your dictionary will enable you to be certain that all nouns you use in your business communications are always correct.

Sometimes a noun error is simply using the wrong word because of an inadequate vocabulary, carelessness, or poor spelling—such as the noun bloopers that follow. These sentences actually appeared in students' compositions.

> The lady put some crullers in her hair.
> The ship was bound for Nausea in the Bahamas.
> My sister writes in her dairy every night.
> When we went to Canada, we had to pass through costumes.

Other types of noun errors occur because English has so many irregular forms—or exceptions. These exceptions, which are especially confusing to foreigners and to young children, can result in errors that are amusing to native English-speaking adults; for example, "My *foots* hurt."

THE MOTHER TONGUE
English, they say, is the language most used,
Most spoken, most written, most cruelly abused.
The plural of box we all know is boxes,
Yet the plural of ox is oxen, not oxes.
A goose is a goose, but two are called geese,
But why isn't more than one moose quoted meese?
A mouse and his family are mentioned as mice,
But the plural of house is houses, not hice.

The plural of brother is brothers or brethren,
And yet we say mothers, but never say methren.
You can readily double a foot and have feet,
But try as you might, you can't make root reet.
Anonymous

Some exceptions, however, are not so obvious, even to people who have used English all their lives; those irregularities are the ones we'll review in this section. The two most important abilities to develop in this section are:

(a) To spot troublesome nouns that are exceptions (irregular).
(b) To interpret the dictionary symbols that enable you to use those nouns correctly.

You'll remember much of what you look up in your dictionary. Therefore, without even trying, you will automatically learn correct forms of troublesome irregulars and improve your spelling and vocabulary.

READ UNIT 19

SAFETY IN NUMBERS

A singular noun means just one of whatever person, place, or thing is named by the noun; a plural noun means more than one. Spelling most plural nouns is easy for anyone whose native language is English. You usually just add **s** to the singular noun to make it plural; for example, *one desk* but *three desks*. When the singular noun ends with **s, z, x, ch**, or **sh**, you add **es** to spell the plural; for example, *one box* but *three boxes*. (See dictionary hint on page 86.)

As you know, however, nearly all English rules have exceptions. Therefore, to be safe when spelling plurals, develop the knack of recognizing nouns that might be exceptions. This unit includes three types of irregular nouns to notice when writing, so that you can verify the plural spelling.

NOUNS ENDING IN Y—YS OR IES?

(a) When a vowel (*a, e, i, o, u*) comes before the **y**, simply add **s** to form plural:

attorn*ey*, attorn*eys* . . . turk*ey*, turk*eys*

(b) When a consonant (all letters other than vowels) precedes the **y**, change the *y* to **i** and add **es**:

company, compan*ies* . . . hobb*y*, hobb*ies*

> It is important to remember that plurals of nouns ending in *y* are often misspelled. Then if you forget the rule, you will remember to refer to your dictionary.

NOUNS ENDING IN O—S OR ES?

(a) When a noun that ends in **o** relates to music, add just an **s** to form the plural:

banj*o*, banj*os* . . . sol*o*, sol*os*

(b) When a noun that ends in *o* does not relate to music, there is no useful rule to determine whether to add *s* or *es* for spelling the plural or whether it's correct both ways:

stud*io*, studi*os* (add *s* only)
tomat*o*, tomat*oes* (add *es*)
domin*o*, domin*os*, domin*oes* (both correct)

> When a plural noun ends in *o*, refer to your dictionary if in doubt—unless it's a music-related word. Should two spellings be shown with a comma or *or* between, both are correct; but it's better to use the first.

NOUNS ENDING IN F—FS OR VES?

There is no useful rule to determine whether to spell the plural of nouns ending in the sound of **f** by:

(a) simply adding *s*

chief, chie*fs* . . . rebu*ff*, rebu*ffs*

or

(b) changing the *f* to *v* and adding *es*

wife, wi*ves* . . . wol*f*, wol*ves*

> Since there is no useful rule, just remember to use your dictionary when spelling the plural of a noun ending in *f* or *fe*.

APOSTROPHES? NO!

In general, apostrophes DO NOT make a noun plural. The occasional exceptions are in Read Unit 48 of Section 8, which is all about apostrophes. Most nouns become plural by adding *S*, not by adding apostrophe *S*. The paragraph below illustrates INCORRECTLY formed plurals. Cross out the apostrophes now.

Wrong Founded in 1926, the Globetrotter's are known worldwide as the *Magician's of Basketball.* Former Secretary of State Henry Kissinger said, "I have particularly admired both the athletic skills's of the Trotter's and the sheer joy and good will they have created over the year's."

Right Founded in 1926, the Globetrotters are known worldwide as the *Magicians of Basketball.* Former Secretary of State Henry Kissinger said, "I have particularly admired both the athletic skills of the Trotters and the sheer joy and good will they have created over the years."—Metromedia, Inc., Annual Report

DICTIONARY HINT FOR PLURALS

In a few dictionaries, the plural spelling is given for every noun that has a plural form. In most dictionaries, the plural spelling is *not* given if all that is needed is to add an *s*—or *es* for nouns ending in *s, z, x, ch,* and *sh*. This means that if you do not find the plural spelling, the noun is not an exception; it is spelled in the regular manner. In a few cases, however, the plural will be shown even when only an *s* needs to be added—when the lexicographer feels that a question might arise, such as with nouns ending in *o*.

REPLAY UNIT 19

PLACE YOUR DICTIONARY BESIDE YOU RIGHT NOW! This is not a spelling test. It is a test of your alertness to words that sometimes present spelling problems and of your ability to use your dictionary efficiently. If you are in doubt about the spelling of any of these words, USE YOUR DICTIONARY. **Do not look at the answers in the back of the book until you have referred to your dictionary whenever you thought it advisable.**

Part A

Spell the plurals of these nouns.

1. ally ______
2. alley ______
3. itinerary ______
4. proxy ______
5. facility ______
6. copy ______
7. journey ______
8. authority ______
9. accessory ______
10. survey ______

Part B

Spell the plurals of these nouns. If there are two correct ways to spell the word, write both in the blank.

1. soprano ______
2. dynamo ______
3. embargo ______
4. piano ______
5. hero ______
6. ego ______
7. potato ______
8. cargo ______
9. portfolio ______
10. memento ______

Part C

Spell the plurals of these nouns.

1. thief ______________
2. handkerchief ______________
3. knife ______________
4. tariff ______________
5. calf ______________
6. half ______________
7. safe ______________
8. wolf ______________
9. plaintiff ______________
10. wife ______________

Part D

Proofread this paragraph for incorrect plurals.

The Globetrotter's are organized into a National and an International team. During a recent tour of the United State's and Canada, they played before more than 1,400,000 fan's in 263 game's. They continue to amuse and delight audience's with a combination of incredible ballhandling, seemingly impossible shooting, and classic comedy routine's.

Answers are on pages 313-314.

DID YOU KNOW THAT . . .
Memos are usually typed on special forms that include the name of the organization at the top and spaces for inserting the following: the name of the receiver, the name of the sender, the date, and the subject of the memo.

Just study the format and read the memo; no other action is required.

VALLEY COLLEGE
INTEROFFICE MEMORANDUM

TO: Coach Musselman, PE Department DATE: April 1, 1984

FROM: Ben Wisenheimer, English Department

SUBJECT: Every Coin Has Two Sides

Remembering our discussions of your football players who were having troubles in English, I have decided to ask you, in turn, for help.

We feel that Paul Barebones, one of our most promising scholars, has a chance for a Rhodes Scholarship, which would be a great thing for him and for our school. Paul has the academic record for this award, but we find that the aspirant is also required to have other excellences and, ideally, should have a good record in athletics. But he does try hard.

We propose that you give some special consideration to Paul as a varsity player, putting him if possible in the backfield of the football team. In this way, we can show a better school record to the committee deciding on the Rhodes Scholarships. We realize that Paul will be a problem on the field; but as you have often said, cooperation between our department and yours is highly desirable. Of course, we do expect Paul to try hard. During intervals of study, we shall coach him as much as we can.

His work in the English club and on the debate team will force him to miss many practices, but we intend to see that he carries an old football around to handle (or whatever one does with a football) during intervals in his work.

We expect Paul to show entire good will in his work for you, and though he will not be able to begin football practice until late in the season, he will finish the season with good attendance.

No offense intended, football players and coaches; we're just having a little fun (at your expense).

READ UNIT 20

PLURALS OUT OF UNIFORM

A plural that doesn't require the addition of an *s* is perhaps like a soldier out of uniform. You cannot recognize his occupation by his appearance, although his behavior might reveal military training. When a noun does not need an *s* to make it plural, it seems to be out of uniform; but its "behavior" in the sentence should reveal its plural status.

Sometimes you recognize a plural without its *s* uniform because of familiarity with the word. To form the plural of these nouns, the spelling is changed; but since they are commonly used, you recognize them instantly:

SINGULAR	PLURAL
man	men
woman	women
mouse	mice
tooth	teeth
child	children
foot	feet

Other plurals that don't require the addition of *s* may be more of a challenge because they are not used as frequently as the words listed above. The nouns that follow have been taken from other languages but are now firmly a part of English. However, the plurals conform to the spelling rules of the language of origin. Sometimes an English spelling has been adopted as well, and then the writer or speaker has two plurals to choose from. In a few cases, the usage of the word changes with the choice of English or foreign plural.

SINGULAR	FOREIGN PLURAL	ENGLISH PLURAL	CHANGE IN MEANING
antenna	antennae	antennas	ae—insects; as—TV
formula	formulae	formulas	ae—scientific, technical
vertebra	vertebrae	vertebras	—
alumnus	alumni	—	—
stimulus	stimuli	—	—
analysis	analyses	—	—
diagnosis	diagnoses	—	—
criterion	criteria	criterions	—
medium	media	mediums	ia—mass communication; ums—communicators between living and dead or intermediary agents, as—mediums of exchange
curriculum	curricula	curriculums	—
datum*	data	—	—
memorandum	memoranda	memorandums	—
addendum	addenda	—	—

*Data is now used by some (but not all) good writers and speakers as both plural and singular.

A well-educated person recognizes nouns that might require foreign plurals and uses the dictionary to be sure of spelling them correctly. Some words seem to fit into the foreign plural category but actually are always used with the regular English form; for example, the plural of *sinus* is *sinuses*, not *sini*.

Care must be taken when you refer to your dictionary to remember that when a plural is "regular," it will probably not be shown in your dictionary. Therefore, if you look up "crisis," you may find next to it "pl. ses" (in some dictionaries). The unsophisticated might fall into the trap and spell the plural "cris**ises**" when it should be "cris**es**." If you were supposed to *add* **es** to form the plural, this would probably not be shown in your dictionary because it would be a "regular" or nonexceptional spelling—like the plural of *boss, bosses*.

Develop an "instinct" for noticing words like those listed on the previous page; remember to refer to your dictionary when using them.

REPLAY UNIT 20

Part A

With the help of your dictionary, spell the plurals of these words. When there are two correct plurals, write them both in the blank.

1. formula ____________
2. addendum ____________
3. alumnus ____________
4. appendix ____________
5. basis ____________
6. bureau ____________
7. census ____________
8. criterion ____________
9. datum ____________
10. index ____________
11. medium ____________
12. goose ____________
13. ox ____________
14. parenthesis ____________
15. diagnosis ____________

Part B

Write *S* or *P* in the blank to indicate whether these nouns are singular or plural.

1. focus ____________
2. stimulus ____________
6. alumna ____________
7. criteria ____________

3. hypotheses ______________ 8. crises ______________

4. vertebra ______________ 9. addenda ______________

5. oxen ______________ 10. basis ______________

You'll find the answers on page 314.

READ UNIT 21

THE ECCENTRIC "S"

An eccentric person or thing is odd or strange and doesn't follow an expected pattern. This unit deals with additional eccentricities of English plurals. Using them correctly will be easy when you've finished this unit.

1. In unit 20 you reviewed words that are spelled differently in the plural form—such as *foot, feet*. Here is a list of words spelled exactly the same whether they are singular or plural. Since you cannot tell by the *s* at the end of the word whether or not it is plural, you rely on other clues when you are reading. In your own writing, be sure to provide clues for the reader. For example, the words *a, one, few*, or *several* might be used as clues:

   A *deer* stopped at the pond. (*Deer* is singular.)

   Several deer lunched on our yellow roses. (*Deer* is plural.)

SINGULAR	PLURAL
deer	deer
sheep	sheep
series	series
statistics	statistics
aircraft	aircraft
salmon	salmon
fish	fish
corps	corps (See the dictionary for pronunciation)
Chinese	Chinese

   Check some of these words in your dictionary; you'll find that each word is respelled following the abbreviation "pl." When a plural is spelled the same way as the singular, the dictionary makes this clear. For some words in the list above, the dictionary shows two correct spellings—either without the *s* or with it. The first is the more commonly seen and heard.

2. Some nouns that end in *s* are usually used in a singular sense. This means that the verb* form that goes with a singular noun is needed.

CIVICS	Civics *was* (not *were*) studied by all the freshmen.
NEWS	The news *is* (not *are*) all bad today.
MEASLES	Measles *is* (not *are*) a contagious disease.
ECONOMICS	Economics *has* (not *have*) been taught here since 1922.

*Further explanation about verbs is found in Section Six.

The pointing word used before the noun is another type of clue that indicates whether a noun is singular or plural:

I like this news. (singular pointing word)

NOT

I like *these* news. (plural pointing word)

3. Some nouns that end in *s* are usually used in a *plural* sense.

trousers scissors clothes goods proceeds

This means that the verb form that goes with a plural noun is used:

The scissors *were* lost.	The goods *have* not yet been shipped.
NOT	**NOT**
The scissors *was* lost.	The goods *has* not yet been shipped.

4. Plurals of proper nouns (ones that begin with capital letters) are spelled in the "regular" way; that is, add *s* to all nouns except those that end in *s, z, x, ch,* and *sh,* to which *es* is added. The spelling of the name should not be changed to form a plural even though it may end in *y, o,* or *f*. For example, if you were to apply the rule given in Unit 17 for spelling plurals of nouns ending in *y* to the proper noun MARY, you would end up with MARIES. To form the plural of *wolf,* you can change the *f* to *v* and add *es—wolves*. However, if you are referring to Mr. and Mrs. Wolf, you should not say "The Wolves are coming over for dinner," but rather "The Wolfs. . . ."

   To make a proper noun plural, just add *s* or *es*; do not add an apostrophe. For example, if I sign my family name to a Christmas card, it will read *The Smiths*, not *The Smith's.*

   In the sample sentences that follow, the underlined words are plural proper nouns:

   The Wolfs have a son who manages the San Diego plant.
   There are two Marys in the Houston office.
   There are many Smiths in the telephone book.
   The Joneses are coming for dinner. (Because **Jones** ends with **s**, add **es** for the plural.)

REPLAY UNIT 21

Part A

Write the plurals of these nouns in the blanks. Use your dictionary when in doubt about the common nouns (nouns that don't begin with capital letters).

1. corps corps	6. Fife s
2. politics Politics	7. aircraft aircraft

3. deer ____________________
4. Mary ____________________
5. series ____________________
8. fish ____________________
9. DeSoto ____________________
10. statistics ____________________

Part B

Write *S* or *P* in the blank to show whether the noun is usually used in a singular or plural sense. Use your dictionary.

1. trousers ____________________
2. economics ____________________
3. measles ____________________
4. clothes ____________________
5. news ____________________

Part C

The verb form in the left column is used with singular nouns, and the verb form on the right, with plurals. Circle the appropriate verb in the sentence.

SINGULAR	PLURAL
is	are
was	were
has	have

1. The trousers (have/has) pockets.
2. Statistics (is/are) my favorite course.
3. These statistics (is/are) accurate.
4. The goods (was/were) shipped yesterday
5. All proceeds from this show (are/is) being given to charity.
6. Clothes (is/are) all over the floor.
7. Measles (is/are) preventable.
8. Each day's news (was/were) carefully edited.
9. A new lecture series (have/has) been completed.
10. Several series (was/were) offered in a variety of fields.

> Did you use your dictionary as well as clues in the wording of the sentences to decide whether each underlined noun subject requires a singular or a plural verb?

Please check your answers on page 314.

Review Units 19–21 before taking POP QUIZ on page 268.

READ UNIT 22

COMPOUNDS AND PROPERS

COMPOUND NOUNS

A noun that is made up of more than one word is called a compound noun—such as *high school*.

COMPOUND NOUNS ARE WRITTEN IN THESE THREE WAYS:

(a) one word checkbook
(b) with hyphen tie-in
(c) two words price tag

No useful rules can help you decide which of the three methods to use when writing a compound noun. Therefore, to be sure of writing a compound correctly, refer to the dictionary. This means that you must develop the ability to recognize an expression that *might* be a compound noun. When you suspect that two or three words might be considered a compound, look at the entry word in the dictionary.

If a dot or an accent mark appears between the parts, the expression is written as one word. The dot or accent mark indicates syllables, not separation of words.

If a space appears between the parts or if the expression cannot be found in a good desk dictionary, each part is written as a separate word.

The noun is a hyphenated expression only if the hyphen appears between the parts of the entry word. (Hyphens are used in expressions other than nouns; these are discussed in Section Ten.)

PLURALS OF COMPOUND NOUNS ARE WRITTEN IN SEVERAL WAYS:

(a) When the compound noun is written as one solid word, we usually spell the plural in the same way as any other noun; that is, add *s* or *es* to the end of the word.

bookshelf—bookshelves spoonful—spoonfuls

(b) When the compound noun is spelled with a hyphen or with a space between the words, the plural form is added to the more or most important word of the compound.

brother-in law—brothers-in-law write-off—write-offs

Since it is not easy to determine accurately which is considered the more important word, it is best to consult your dictionary.

PROPER NOUNS

A noun that begins with a capital letter is called a proper noun. (Nouns that do not begin with capitals are common nouns.) A complete list of capitalization rules is in the back of this book. In this unit you review only those principles of noun capitalization that are most often questioned by business writers.

1. Official titles are capitalized only when used *directly* before the name of the person:

I think **P**resident Humphrey Bogart called the meeting to order.
Give the awards to **C**olonel Alfred Adler.
A letter of recommendation was received from **S**ister Mary Therese.

BUT

Humphrey Bogart is the **p**resident of the company.
The **c**olonel, Alfred Adler, is a graduate of West Point.
(No capital for *colonel* because a comma separates it from the name.)
Rachel Rothstein, a **p**rofessor of business, sent the letter.

Exceptions (a) Titles of *high*-ranking government or church officials are always capitalized when they refer to a specific person and not just anyone of that rank:

The **P**ope is going to visit the United States this month.
Harry S. Truman was **P**resident at that time.

(b) Titles that are used in addresses and signatures are capitalized even when they appear after the name:

(*In address*) Mr. Frank Carter, General Manager
(*In signature*) Frank Carter, General Manager
(*Mr.* shouldn't be used in signature.)

2. Words such as *company, college,* and *association* are usually capitalized only when they are used with the name of the organization:

Tully Plumbing **C**ompany is on Fifth Street.

BUT

The **c**ompany is on Fifth Street.

3. Capitalize the name of a department or a committee only when you are a member of the organization:

Make out a requisition and send it to **P**urchasing.

OR

Make out a requisition and send it to the **P**urchasing **D**epartment.

BUT

He works in the **p**urchasing **d**epartment of our major competitor.

4. Capitalize the words *town, city, state, county,* and so on only when they come after the name:

New York **S**tate, New York **C**ity, and Kansas **C**ity

BUT

the **s**tate of California and the **c**ity of Portland

5. Capitalize names of geographic sections that have compass point names, but do not capitalize directions:

He lives in the **E**ast, but he talks like a **W**esterner.
Disneyland is **e**ast of Los Angeles.

6. Do not capitalize the names of seasons:

The office will close for a one-week vacation this **s**ummer.

7. Always capitalize names of languages:

English, **F**rench, **S**panish, **H**indi, **S**wahili

REPLAY UNIT 22

Part A

Refer to your dictionary to determine how these compound nouns are spelled. Show whether the expression is one solid word, a hyphenated word, or two or three words. Some dictionaries show more than one spelling for some of the plurals; if so, include both.

Example		SINGULAR	PLURAL
	notary public	notary public	notaries public notary publics
	1. spoonful	Spoonful	S
	2. textbook	text-book	S
	3. microprocessor	micro-processor	S
	4. editorinchief	Editor-in-chief	Editors in chief
	5. stockholder	Stock-holder	S
	6. passerby	Passer-by	Passers-by
	7. businesswoman	Business-woman	Businesswomen
	8. brotherinlaw	brother-in-law	Brothers-in-Law
	9. timetable	time-table	S
	10. billoflading	bill-of-lading	Bills-of-lading

COMPOUND SPORTS

When the games of basketball, baseball, and football were first invented, they were written as separate words; basket ball, foot ball, and base ball. As each game became more popular, it became a hyphenated word: basket-ball, foot-ball, base-ball. Eventually all became the one-word compounds that they are today.

Part B

If the capitalization in the sentence is correct, write C in the blank. Otherwise make the necessary corrections. Refer to the information in this unit as well as to "Capitalization" in the the back of the book.

Example If their ~~S~~ales ~~M~~anager calls, let me talk to her. (s, m)

1. Willy Mays is the General Manager of Coronet Manufacturing Company. ______
2. We make it a policy to hire former professional athletes to work in the Sporting Goods Department. ______
3. Who is the Manager of their Shipping Department? ______
4. The association relies on its members for distributing the information. ______
5. The secretary of state has just entered the White House. ______
6. Do you think Cardinal Manning discussed the problem with the bishops? ______
7. The typed signature at the end of the letter should appear like this: Marianne Atherton, office manager. ______
8. I headed North last Spring, calling on every appliance dealer between here and Carson city. ______
9. The City of Azusa, which is in the State of California, is named after everything from A to Z in the USA. ______
10. The clerks in our Credit Department speak english and spanish. ______
11. My Uncle has taught Business English at several Colleges. ______
12. Until the 1950s, there were no Black or Jewish secretaries in this Company. ______
13. We bought Hewlett-Packard computers for all our offices west of the Mississippi River. ______
14. Governor Shawn A. Taylor joined his famous sisters, Christa and Ashley, at the Inauguration Ball. ______
15. Christa Taylor won the academy award for her performance in *The Iron Magnolia*, and Ashley Taylor was awarded the nobel prize for her efforts on behalf of world peace. ______

The answers are on page 315.

READ UNIT 23

UNISEXING ENGLISH FOR CAREERS

Some of the fallout from the women's movement of the 1970s included unisex hairstyles and clothes. Of concern to us in improving English for careers is the concept of unisexing English. We don't want to offend some of our customers or clients, and we do want our writing and conversation to make us appear knowledgeable about today's realities.

As with other movements throughout history, the pendulum may swing too far in the opposite direction before a balance is achieved. Thus we have the traditional "sexist" expressions that are currently still in use because of habit and thoughtlessness. We also have the reverse: self-conscious expressions that draw attention to themselves instead of to the message.

Professionals and business people need to sense the words that may offend, the expressions that may arouse ridicule, and the words that language authorities feel detract from the grace of the English language. I'm sorry I can't provide you with hard-and-fast rules. Who can establish rules for what amounts to good taste? I can, however, provide you with some do's and don'ts for specific types of words.

This unit is designed to remind you to be thoughtful of the sexism issue in oral and written communications related to your career. Your own good judgment plus an alertness to how others are avoiding the traditional sexist overtones in their language will enable you to apply these concepts.

AVOID	INSTEAD, USE
girl, girls (when referring to adults) lady, ladies (when used where *woman* or *women* would sound more natural)	woman, women, or other identifying term, such as occupation of the female

AVOID	The ladies in my office dress better than the men.
AVOID	The ladies in my office dress better than the gentlemen.
USE	The women in my office dress better than the men.
USE	She is truly a lady in every sense of the word.
USE	The ladies of the Court bowed when the King and Queen entered.

AVOID	USE
authoress	author
poetess	poet
directress	director
Negress (interpreted as a symbol of both sexism and racism)	black woman or black girl, depending on age, or omit the color reference
Jewess (interpreted as a symbol of both sexism and mild anti-Semitism)	Jewish woman or Jewish girl, depending on age, or omit the reference to religion

> Nouns such as *waitress* and *actress* have been in general use for a long time and continue to be in good taste. Use your judgment to determine whether feminization with *ess* is appropriate.

AVOID	USE
women's libber or libber	feminist
women's lib	women's movement
career girl, career woman	name the occupation; have you ever heard of a career boy or a career man?
manmade	synthetic, artificial, manufactured
manpower	workers, work force, human resources
chairman (OK)	head, chair, leader, or moderator is preferred rather than chairwoman or chairperson
businessmen	executives, managers, entrepreneurs, business people
workmen	workers, laborers, crew, staff
cameraman	photographer
male nurse	nurse
male secretary	secretary
lady doctor	doctor
woman manager	manager
lady lawyer	lawyer
man and wife	man and woman or husband and wife
men and girls	boys and girls or men and women
mailman	letter carrier
fireman	firefighter
policeman	police officer

In an attempt to avoid sexism in language, such clumsy expressions as "Gentlepeople" or "Gentlepersons" as the salutation of a business letter and "personhole" to replace "manhole" have been suggested. Language authorities feel such expressions sound too artificial and believe their use is an insult to English.

The term "sexism" originally referred to discrimination based on sex (just as "racism" referred to discrimination based on race). However, it now refers more often to thoughtless stereotyping of people based on their sex. Be alert to unisexing your language, but don't let unisex result in clumsy wording. Use your judgment.

For further suggestions, see *WITHOUT BIAS A Guidebook for Nondiscriminatory Communication* by Judy Pickens, Ed., 2nd Edition. Business communicators refer to this handbook for help in avoiding race, sex, age, and disability biases in language.

REPLAY UNIT 23

Update the language in these sentences so that they will reflect today's attitudes instead of yesterday's stereotypes.

1. The girls in my office go to lunch at 12.
2. He promised to send his girl over with the contracts.
3. The directress of the organization is a Negress.
4. Are you one of those women's libbers?
5. After dinner the men discussed the election while the girls had their committee meeting.
6. We have several lady policemen guarding against intruders.
7. He is studying to be a male nurse.
8. There's a lady lawyer in the office across the hall.
9. The authoress of the book you are now reading doesn't like to be called an "authoress."
10. The chairman of the committee is a career woman.

See page 315 for answers.

DID YOU KNOW THAT . . .

- The word *gymnasium* entered the English language through the Greek word "gymnos," which means "naked." The ancient Greek athletes trained naked.
- The word *curriculum* originally meant a race course in Latin—and there are still some today who believe that a curriculum is a runaround.
- The word *school* was derived from the Greek word "skhole," which means "leisure" or "spare time," since it was only the leisure classes that could afford an education.
- The study of the origin of words is called *etymology*.

READ UNIT 24

VOCABULARY AND SPELLING

Good vocabulary is vital to obtaining and holding a good job in business. Vocabulary building involves learning new words and adding meanings to words previously acquired. Here are the steps usually followed, although we are often not consciously aware of them as they take place:

1. We recognize a written word; that is, it looks familiar to us.
2. Eventually we understand the meaning of that word when we **read** it because we have seen it a few times within sentences.

3. Later we understand the meaning when we **hear** the word.
4. After that, we are ready to use the word in our own writing.
5. Finally, we're able to use the word correctly in conversation; this means that the word has been mastered.

The process just described continues as long as we read a wide assortment of materials, watch movies and TV with alertness, and listen to and converse with a variety of people. The result is a gradual increase in vocabulary over a period of years. However, this natural process is slow, and we can speed up vocabulary growth from time to time by consciously working on it. This section may include some words that you have not yet mastered. Now would be an appropriate time to increase your vocabulary by learning them.

An educated person is expected not only to have an extensive vocabulary but also to know how to spell. While studying the vocabulary in this unit, LOOK carefully at the words. You'll find you will know how to spell most of them by the time you've finished the unit. Even any you're not sure of will look "funny" if misspelled, prompting you to refer to a dictionary for a correct spelling.

When studying any subject, you can improve spelling and vocabulary by learning words used in the textbook, the lectures, or the discussions. This unit provides the opportunity to master words that appeared in this section of the text. Here are sentences in which words from Units 19–22 are used. Each word for you to take note of is underlined.

> Some words selected for these sentences present spelling difficulties only; others might require the use of a dictionary to verify meanings. Review both singular and plural spellings as well as definitions.

1. My allies left the cargo in the alley.
2. Please put the bill of lading in the portfolio, as it will be a good tie-in with the evidence.
3. She was given a copy of the itinerary as a memento.
4. Never before in this school's history has an alumna broken her vertebrae by lifting a chassis.
5. What criteria are you using to determine the amount of the tariff?
6. There is an embargo on dynamos from Pandora.
7. The plaintiff insists that the proceeds from the sale of the factory should be distributed equally to the partners.
8. Several new hypotheses are included in the addendum.
9. The survey of the alumni was completed last week.
10. Proxies from 1,000 stockholders were received at this facility.
11. Some Native American tribes used shells and blankets as mediums of exchange
12. When my proposal was rebuffed, I accused the bureau chief of bias.
13. The new curriculums were widely advertised in the local media.
14. The runners-up in the sales contest each received a marketing textbook.
15. The accountant recommended using the loss as a tax write-off.

REPLAY UNIT 24

Check your spelling and vocabulary prowess.* All the words below are in Section 4.

1. A schedule listing departure and arrival time for trains or buses. time-Table
2. My husband's sisters sisters-in-law
3. Several male graduates or male and female graduates alumni
4. One who starts a legal action plaintiff
5. Standard on which a decision can be based criterion
6. Plural of answer to No. 5 criteria
7. People working together for a cause allies
8. Narrow road behind buildings alley
9. Freight carried by a vehicle, especially a ship cargo
10. Receipt from a shipping company bill of lading
11. Carrying case for holding papers; a list of investments portfolio
12. An elimination of an item from an account write-off
13. Profits from a commercial or other venture proceeds
14. Travel plan, including transportation, times, dates, hotels, etc. itinerary
15. Object that is reminder of past, such as a souvenir memento
16. Frame for a vehicle chassis
17. Plural of answer to No. 16 chassis
18. Written authorization to act for another proxy
19. Belief that is stated for purpose of experimenting, arguing, reasoning, or coming to tentative conclusions hypothesis
20. Government order prohibiting trade with a particular country or of a product embargo

Check your answers on page 315.
Now take POP QUIZ on page 269.

Checkpoint

A *regular* noun becomes plural by adding **s**—or by adding **es** for those ending in *s, x, z, sh,* and *ch.* Many nouns, however, don't follow the preceding rule; that is, they are *irregular.* You've reviewed in this section irregular

*You expected to find the definition down here? Look it up.

plural spellings and pronunciations as well as the dictionary symbols for plural forms.

In addition to irregular plurals, business writers need expertise with compound nouns, capitalization, and avoidance of sex-biased language.

While reviewing Section 4, you simultaneously increase your vocabulary and spelling ability, as emphasized in Unit 24.

WORD PLAY The words represented by the following brief definitions rhyme with BIRD. Fill in the blanks.

1. Listened ______
2. You should put one in each blank ______
3. A group of large animals ______
4. Unreasonable and ridiculous ______
5. Next after second ______
6. The thick part of sour milk ______
7. Went astray or made a mistake ______
8. Postponed ______
9. Agreed; were of the same opinion ______
10. Came together to discuss ______

If you want a hint for any of them, here's the first letter of each:

1. h	6. c
2. w	7. e
3. h	8. d
4. a	9. c
5. t	10. c

Special Assignment

In the Special Assignment for Section One, you were encouraged to write CLEARLY. The ability to write CONCISELY is also vital to successful business communication. This assignment provides practice in writing concisely as well as clearly. The subject is one that could be used for a long essay in an English composition class. Only 100 to 120 words, however, will be acceptable for this course.

Write about any job you have held. You might include some or all of the following information in *paragraph* form. (If you've never had a job, write about one you wish you had.)

kind of work you did or do	approximate dates of employment
name of company	part time or full time
type of business	what you liked or didn't like
name and/or title of supervisor	something you learned

Have your dictionary close by as you work. Use Standard English—no slang. You will have several opportunities to apply rules for correct capitalization.

Here are some sample opening sentences:

1. Selling tires at Sears during the summer of 198- was a valuable experience for me.
2. I was a part-time typist for the Fitrite Shoe Company during my senior year in high school.
3. During the summer of 198-, I assembled lamps for Aurora Lighting Company.
4. Working under the supervision of Ms. Bush was an experience I would never want to have again.
5. My first job after completing college was as a computer programmer for a small electronics firm.

Sometimes I receive assignments from students who stretch out their writing as one might do with a piece of taffy. When that happens, the ideas that should be combined into one concise sentence appear in several rambling sentences with many unnecessary words. Compare the following with Sample Sentence 1:

> I am going to tell you about my job. The work I did on my job was selling. I was a tire salesman. I sold tires at Sears. I did this work during the summer. It was two years ago in 198-. The experience I had was valuable to me.

After composing the opening sentence, write several more sentences giving some details about the work. If you sold tires at Sears, you might tell about how difficult it is to deal with the public and some techniques you used to handle customers effectively.

The closing sentence should be a summary or conclusion.

> If possible, type this assignment and use double spacing.
> This report is due ____________ (date).

Proofreading for Careers

Proofreading is a challenge because it requires concentration on several subjects at the same time:

> spelling, word division, punctuation, grammar, format, typographical errors, sentence construction, capitalization, word choice, number accuracy, number usage, meaning

With your dictionary beside you, use your pen held upside down or closed to "underline" slowly each line of type. The "underlining" changes to pointing when you look at a possible error. This underlining-pointing process focuses your attention on proofreading.

Circle and correct spelling, typographical, sentence construction, and capitalization errors in the letter to Ms. Gonzalez:

Dear Ms. Zonzalez:

We can help your Sales Staff by providing you firm with a new demension of client service—an interior design tiein with the sale of office space.

Clients looking for office space often ask about desks, filling cabinets, chairs, and carpeting. They are interested the in prices and availability of these items, this is were we can come in to help close the sale for you.

At not charge toyou, we will furnish a complete plan that will be custom-designed to fit any size office area that you offer. This will be an importent sales aid to you that will pay off in faster and increased sales.

Alert Realty Firms like yours are always looking for new and creative consepts in selling. The next time you have a cleint with an office furnishing need. Give us a call or reffer your client to us, we will proof that this service will work for you.

We are specialist's in office interior design and show great versatility in meeting individual and general needs. Are designer's are available to assist you or your client in selecting major items or decorator accessorys.

May get to gether with you or one of your key sales representative's soon so that we can discuss presise arrangemnts we can make with you.

Yours Very Truely,

Manny Errata

What's your score? Excellent? Good? Fair? or?? ______________

For proofreading expertise, complete the "Reference Manual" Reads and Replays on pages 281–293.

Practice Quiz

Read each sentence carefully to determine which form makes sense in that sentence and is spelled correctly. You may use your dictionary; however, know the material well enough so that you can easily finish within 20 minutes.

1. These (a) phenomena (b) phenomenas (c) phenomenae (d) phenomenon were discovered (Unit 20)
2. through the marketing (a) survey's (b) survies (c) surveys. (Unit 19)
3. The (a) stock-holders (b) stocks holder (c) stockholders (d) stock holders (Unit 22)
4. will mail in their (a) proxys (b) proxi (c) proxyes (d) proxies. (Unit 19)
5. Several (a) beneficiary's (b) beneficiarys (c) beneficiaries are (Unit 19)
6. (a) allies (b) allys (c) alleys (d) alloys in this controversy. (Unit 19)
7. His (a) brother-in-laws (b) brother in laws (c) brothers-in-law (d) brothers-in laws manage the office. (Unit 22)
8. The manager wants all employees to learn the (a) cargos (b) criteria (c) proceeds (d) mementos for requesting printouts.
9. We have two accounts of enemy-downed (a) aircraft (b) aircrafts (c) air-crafts (d) air craft. (Unit 22)
10. The ranchers told (a) Congress man (b) congressman (c) Congressman (d) Congress-man Lane of their problems with their (Unit 22)
11. (a) sheeps and oxes (b) sheep and oxen (c) sheeps and oxen (d) sheep and oxes. (Units 20 and 21)
12. The nouns (a) cargo and embargo (b) vertebra and chassis (c) addenda and appendix (d) tariff and bill of lading have almost the same meaning. (Unit 24)
13. Both (a) attornies (b) attorneys (c) attornys (d) attorney's seemed confident as they awaited the verdict. (Unit 19)
14. The inventory indicates that we have only three (a) celloes (b) cello (c) cellos in our Music Department. (Unit 19)
15. Although James McCarthy is the president, three other (a) McCarthy's (b) McCarthies (c) McCarthys (d) McCarthys' hold managerial positions (Unit 21)
16. with this (a) company (b) Company. (Unit 22 and page 285)
17. The medical students submitted accurate (a) diagnosis (b) diagnosises (c) diagnoses (d) diagnosi. (Unit 20)

b 18. The (a) Jones's (b) Joneses (c) Jones (d) Jones' invited the entire staff to dinner. (Unit 21)

b 19. Our (a) Marketing department (b) Marketing Department (c) marketing department (d) marketing Department (Unit 22 and page 285)

a 20. is headed by the (a) merchandise manager, (b) Merchandise manager, (c) Merchandise Manager, (d) merchandise Manager, Mr. Frank Perez. (Unit 22)

a 21. We will be closed for two weeks during the (a) summer (b) Summer. (Unit 22)

a 22. A (a) congresswoman and a waitress (b) female congressman and lady waiter (c) woman congressman and a waitress (d) congressman and waitress were both trying on the same style of dress at Marshall Field's. (Unit 23)

b 23. The (a) senator (b) Senator from Oregon voted for a million dollar grant to Clatsop Community College. (Unit 22 and page 284)

D 24. (a) The men will take their girls with them to take the notes. (b) The men will take the ladies with them to take the notes. (c) The gentlemen will take their secretaries with them to take the notes. (d) The managers will take their secretaries with them to take the notes. (e) The girls will go along to take notes for the men. (Unit 23)

e 25. Mr. Campbell's wife is a (a) female policeman (b) female police officer (c) lady policeman (d) woman policeman (e) police officer. (Unit 23)

SECTION 5

BE KIND TO THE SUBSTITUTE WEEK

After Completing This Section, You Will:

- *Use pronouns according to the principles of Standard English that are acceptable to most successful professionals and business people today.*

Some people delight in reminiscing about junior high school experiences with substitute teachers. The appearance of a substitute often was the signal for class buffoons to go into action, and even usually well-behaved students would sometimes join in the fun. When the substitute began to cry, the students knew they had achieved the ultimate in success. As adults, you probably feel a bit remorseful now when you recall how mean your classmates (never *you*, of course) were.

This chapter gives you a chance to be kind to substitutes—noun substitutes, that is—by using them correctly. A word that SUBSTITUTES for a noun is called a **pronoun**. A pronoun is a word that refers to a person or thing that was previously identified by a noun. For example, after identifying a thing as a "pencil," you may later use the pronoun "it"; the reader or listener will understand that "it" means *pencil*. To substitute for the plural noun "pencils," the pronoun "them" might be used:

Give the *pencils* to me. I need *them* now.

Several pronoun rules that used to be essential elements of Standard English are now being violated by prestigious people in business and professions. For example, according to traditional grammar textbooks and English language purists, this is what the dialogue should be like when you knock on someone's door:

Host: Who's there?
Visitor: It is I.

The use of "I" in the preceding conversation is technically correct, and many well-educated people do follow the traditional pronoun rules. However, a number of today's English language authorities would recommend the above reply be changed to "It's me." The rule in question is in transition, and that is why the authorities don't agree.

That's also why the rule, invented in the 18th century, involving the use of the nominative case after existence verbs will not be explained in this section. In fact, the only place you'll find information about the nominative case or any other case is in the Grammar Supplement in the back of the book.

Another problem area has been distinguishing between *who* and *whom*. Have you ever heard someone correcting himself like this?

"Who ah whom ah who ah is the treasurer of the company?

Some who/whom cowards mumble when the mystery word is needed in the middle of a sentence and hope the listener will not notice. Modern language authorities have simplified the dilemma, as you will read later in this section.

Just learn to apply the basic and easy principles given here. Then you'll be sure your pronoun usage is suitable for nearly all communications.

READ UNIT 25

JUST BETWEEN YOU AND ME

Or should it be "Just Between You and I"? Are you sure? Well, let's settle it by saying "Just Between Us." There's no question about that; it couldn't be "Just Between We."

Notice that as soon as you replace the two pronouns with one, the wrong answer "sounds" wrong. However, with **two** pronouns—you and me/you and I—it's harder to tell which one belongs.

Which is right?
John and I? or John and me?
him and her? or he and she? or he and her?

Only one of the preceding combinations is wrong—the last. You can't tell about the others until they are used in sentences.

Most pronoun errors occur when two or more are used together or when the name of a person is used with a pronoun. You will usually make the appropriate choice if you imagine one of the two words left out and then decide whether the sentence "sounds right."

John and me went to the movies. TERRIBLE!

Leave John home, and you will immediately know that "Me went to the movies" doesn't sound right. You would therefore change the sentence to

John and I went to the movies. GOOD
Give the papers to John and I at once. HORRIBLE!

Leave out John, and you will immediately know that "Give the papers to I at once" doesn't sound right. You would therefore change the sentence to

Give the papers to John and me at once. GOOD

. . . . . .

Sometimes the noun and pronoun are used together without a conjunction between them. This happens when a noun and pronoun that mean the same thing are used together for emphasis or clearness.

Which is right?
we secretaries? or us secretaries?
we boys? or us boys?

The preceding combinations are all Standard English. You can't tell which form to use until you see the rest of the sentence. You will usually make the appropriate choice if you imagine the **noun** left out and then decide whether the sentence "sounds right."

Us secretaries need a longer lunch period. AWFUL!

Omit the noun *secretaries* and you will immediately know that "Us need a longer lunch period" doesn't sound right. You would therefore change the sentence to

We secretaries need a longer lunch period. GOOD
Will you please give we secretaries a longer lunch period? DREADFUL

Omit the noun *secretaries* and you will immediately know that "Will you please give we a longer lunch period" doesn't sound right. You would therefore change the sentence to

Will you please give us secretaries a longer lunch period? GOOD

. . . . . .

If applying the trick methods shown above seems awkward in a particular sentence, refer to this more formal explanation. The **Subjects** and **Objects** columns below show the personal pronouns that are sometimes confused when used with another pronoun or with a noun. The ones on the left always fit as the *subject* of a sentence. The ones on the right always fit as an *object* of a preposition or of a verb. For an easy test, just remember this sentence:

I love him.

Any pronoun that fits at the beginning of the preceding sentence will be a "subject" pronoun. If it fits at the end, it is an "object" pronoun.

	SUBJECTS	OBJECTS
1st person	I, we	me, us
2nd person	you	you, (You can't go wrong since they're the same.)
3rd person	he, she, they, it	her, him, them

You can now answer the question in the opening paragraph of this unit. Since "Just between *us*" sounds right, a pronoun in the Objects column above is required; therefore, "Just between you and *me*" is correct. The **object** pronoun *me* is used instead of the **subject** pronoun *I*. The pronoun *me* is the "object" of *between*, a preposition.

REPLAY UNIT 25

Use the simple methods described in this unit to determine the correct pronoun; then underline it.

1. If you assign the report to John and (I, me), it will never be completed.
2. Please tell Anna and (I, me) when you are ready to leave.
3. The supervisor and (he, him) are working on the project.
4. Please give more homework to (we, us) students.
5. You and (I, me) should work together frequently.
6. (We, Us) shipping clerks should affiliate with the union.
7. Mr. Gomez telephoned Mr. Potter and (I, me).
8. The manager is sorry that she assigned the work to you and (him, he).
9. Do (we, us) parents get a chance to vote?
10. You and (her, she) should plan to take the trip together.
11. Mrs. Hamilton gave Kenneth and (I, me) the books.
12. Peggy asked you and (we, us) to visit her.
13. (They, Them), as well as Mr. Scott, will serve on the committee.
14. Everyone except Mr. Jones and (he, him) will be at the San Francisco session.*
15. (Him and me, He and I) have an advantage in this situation.
16. The money should be divided between you and (I, me).
17. Mr. Heffron saw you and (she, her) arrive late to class.
18. (We, Us) accountants can no longer work under these conditions.
19. Please allow Ms. Fisher and (I, me) to attend the conference.
20. All the responsibility was given to the auditor and (I, me).
21. The personnel director told Sean and (I, me) about the job requirements.
22. Were you and (he, him) using the Dictaphone that day?
23. Mr. Hicks invited you as well as (he, him).
24. The production manager doesn't want Barry and (I, me) to work together.
25. You and (them, they) should devise a new production schedule.

Please check your answers on page 316.

*Subject is **everyone**; **except** is a preposition.

READ UNIT 26

ME, MYSELF, AND I

When you were a child, you probably asked your mother questions like "Can Johnny and me go to the movies?" Your mother may have replied "Johnny and *I*" and wouldn't give you the money until you changed that evil word "me" to "I." Although she was right, a number of incidents like that through your years of growing up may have made you a firm disbeliever in the word "me." It really is quite a respectable word and does not deserve an X rating, even though some people when trying their best to be correct will try to avoid "me." In this unit we'll practice making additional choices between various pronouns.

COMPARISONS

When the Unit 25 method doesn't apply, decide whether there are some "understood" words that could complete the thought. When "understood completion" words are helpful, the sentence usually involves a comparison—that is, comparing somebody with someone else.

(a) No one wants to please you more than the manager and (I, me).
If you add the "understood" word *do* after the questionable pronoun, you can easily decide to use *I* instead of *me*.

No one wants to please you more than the manager and I (do).

(b) Do you like him better than (I, me)?
If you add the "understood" word *do* after the questionable pronoun, you can easily decide to use *I* instead of *me*.

Do you like him better than I (do)?

BUT

Selection of the pronoun can determine the meaning.

Do you like him better than (I, me)?

If you add the "understood words *you like* **before** the questionable pronoun, you can easily decide to use *me* instead of *I*.

Do you like him better than (you like) me?

(c) Although he is older than (she, her), he is shorter.
If you add the "understood" word *is* after the questionable pronoun, you can easily decide to use *she* instead of *her*.

Although he is older than she (is), he is shorter.

(d) It doesn't surprise me as much as him. (See this example in *American Heritage Dictionary* under **than**.)

It doesn't surprise me as much as (it does) him.

By thinking the understood words *it does*, you recognize that *he* would be wrong.

SELF, SELVES

When *self* or *selves* is added to a pronoun, the resulting word is called a compound pronoun. Here are the Standard English compound personal pronouns. Any other spellings of these words are unacceptable in business and professional communications.

myself	itself
yourself, yourselves	oneself
himself	ourselves
herself	themselves

Careful writers use compound pronouns only when **no other pronoun with the same meaning will make sense in the sentence**. The "self" pronouns are used either for emphasis or for clearness.

A compound pronoun should match a noun or pronoun already in the sentence:

We ourselves are to blame. (for emphasis)
He corrected **himself** immediately. (for clearness)

AVOID The manager and myself will attend the meeting.
A careful writer would not use *myself*, because *I* makes sense.

USE The manager and I will attend the meeting.
If you leave out "the manager," as explained in Unit 25, you'll know that *me* would be incorrect.

USE She does the easy work by herself.
No other pronoun would make sense after "by"; therefore, *herself* is correctly used.

USE She herself does the easy work.
No other pronoun would make sense after "she"; therefore, *herself* is correctly used.

Here is the test for compound pronouns. Can you insert another pronoun at the exact place where the "self" or "selves" word appears? If so, use the other pronoun. If not, keep the "self" or "selves" word right where it is, but be sure that it is correctly spelled.

WHICH STATEMENT IS GROUNDS FOR DIVORCE?
- My husband likes golf better than I.
- My husband likes golf better than me.

REPLAY UNIT 26

Part A

Make the necessary pronoun corrections. Write *C* to the left of the sentences that are already correct.

______ 1. The president himself will attend that meeting.

______ 2. The twins learned how to dress theirselves at an early age.

______ 3. She can do the work faster than him.

______ 4. When it comes to his work, he is more confident than she.

______ 5. Give the papers to Mr. O'Connor and myself.

______ 6. His younger brother is just as tall as him.

______ 7. The other workers always leave as soon as myself.

______ 8. Mr. Adams does not work as efficiently as she.

______ 9. Since he is a physician, he gave hisself the injection.

______ 10. No one wants to please you more than the manager and myself.

______ 11. The lawyer and myself listened to the case with great interest.

______ 12. We may find ourself looking for a new manager.

______ 13. We ourselves are excluded from the contract.

______ 14. The members of the cast felt like themselves again after the crisis had passed.

______ 15. She injured herself.

______ 16. We voted ourself a pay increase.

______ 17. They voted theirself a pay increase.

______ 18. He voted hisself a pay increase.

______ 19. I voted myself a pay increase.

______ 20. You voted yourselves a pay increase.

Part B

Show by the "understood words" method why all these sentences are **correct**. Fill in the parentheses with the understood word or words.

1. He loves his wife more than she.

   He loves his wife more than she (______).

2. He loves his wife more than her.

   He loves his wife more than (______) her.

3. I know the vice president better than he.

   I know the vice president better than he (______).

4. I know the vice president better than him.

   I know the vice president better than (______) him.

5. Ms. Mosson can operate the device as well as he (______).

Check your answers on page 316.

READ UNIT 27

THE LIZARD'S TAIL

If, as a child, you never tried to pull a lizard, tail first, out of a crevice in a rock, you've missed the shock of a lifetime—that of discovering that although its tail was in your hand, the rest of it didn't come along. Later on you learned that this was not as horrible as it seemed, for the lizard managed to grow a new one.

When you wanted to tell your buddy, you said in amazement, "The lizard lost its tail!" Although you didn't know it at the time, you were using a possessive pronoun as a substitute for a possessive noun. If you hadn't substituted, you would have said, "The lizard lost the lizard's tail," using the possessive noun* and sounding rather silly.

The following sentences illustrate how possessive pronouns may substitute for possessive nouns, such as "Joe's."

That is *my* book.	That book is *mine.*
That is *his* book.	That book is *his.*
That is *her* book.	That book is *hers.*
That is *our* book.	That book is *ours.*
That is *your* book.	That book is *yours.*
That is *their* book.	That book is *theirs.*
The lizard lost *its* tail.	*Whose* tail is that?

Notice that all the possessive pronouns in the preceding sentences are NEVER used with apostrophes.

Some *possessive pronouns* and *contractions* sound alike, but they are spelled differently. In these expressions below, the **contractions** require **apostrophes**. The apostrophe is used to replace the space and letter that have been left out to form the contraction. Apostrophes are NOT used with the possessive pronouns (except with "one" and "body" pronouns, as shown on the following page).

POSSESSIVE PRONOUNS	CONTRACTIONS
your—Your job is interesting.	you're (you are)—You're a good typist.
its—Its wing was injured.	it's (it is)—It's a good machine.
whose—Whose book is on the desk?	who's (who is)—Who's going to work?
their—Their problem is serious.	they're (they are)—They're in trouble.

*Possessive nouns are presented in Section Eight.

Here are some more pronouns. To form possessives of these pronouns, add **'s**:

PRONOUNS	POSSESSIVE PRONOUNS UNDERLINED
anybody	That could be anybody's calliope.
somebody	It must be somebody's saxophone.
everybody	That is everybody's banjo.
anyone	That isn't anyone's business.
someone	Someone's harp was left behind.
everyone	Everyone's books are on the table.
one	One should mind one's own business.

The "body" and "one" words listed above may also be used as part of contractions:

Everyone is	Everyone's going to the concert.
Somebody is	Somebody's in the kitchen with Dinah.

REPLAY UNIT 27

Correct the pronoun and contraction errors in the following sentences. Write C to the left of the sentence if it is already correct.

____ 1. Are you the one whose going to attend the sales meeting this year?

____ 2. Whose work do you prefer?

____ 3. It's going to be difficult to prepare this report.

____ 4. Its color has faded.

____ 5. The carpeted office is our's.

____ 6. Your's is on the 18th floor.

____ 7. I think the one on the right is hers.

____ 8. The papers are mine; the books are their's.

____ 9. Your to use your own books today.

____ 10. I don't believe your's is here.

____ 11. Their all going with us.

____ 12. No ones work was checked yet.

____ 13. Whether or not we'll get that account is anybody's guess.

____ 14. Whose going to prepare your format?

____ 15. Almost anyones work would have been better than their's.

Answers to this Replay are on page 316.

Review Units 25–27; then turn to Pop Quiz on page 270.

> **DID YOU KNOW THAT**
> STUDENT
> is a word derived from the Latin "studium" meaning "zeal" or "eagerness."
> SOPHOMORE
> literally means "wise fool." It is an invented word combining two Greek words: sophos (wise) and moros (foolish)
> PEDAGOGUE
> which now means "teacher," originally meant a slave in charge of children—from the Greek "paidagogos." The Romans, who respected the learning of the Greeks, assigned Greek slaves to train their children.

READ UNIT 28

TWO'S COMPANY; THREE'S A CROWD

That's what the girl whose boyfriend was visiting used to tell her little brother. The boyfriend would then toss a quarter to the brother and suggest he go to the movies. That's the way it was in the good old days.

A crowd, however, still means a collection of people; and "crowd" is a sample of a COLLECTIVE NOUN. Collective nouns are words that represent a group of people or animals. Notice that each collective noun below is singular, even though it represents a collection of people or animals.

class, company, jury, herd, committee, team, staff,
group, union, family, or the name of any organization

Pronouns frequently substitute for collective nouns. Otherwise we'd have awkward repetition of the nouns, as in the following:

The company closed *the company's* offices for the month of July.

In this sentence a pronoun can substitute for the repeated noun. Singular pronouns such as *it, its,* or *itself* are used to substitute for collective nouns when the members of the group are acting as a unit—or as one. For example, even though the company has hundreds of employees, the idea of *one* company is being expressed. Therefore, when choosing the pronoun, be sure it is singular:

USE The company closed **its** offices for the month of July.
AVOID The company closed **their** offices for the month of July.

Occasionally, however, a sentence is about the members of a group who are acting as separate individuals or disagreeing. Then use a plural pronoun: they, their, them, themselves.

USE The jury had **their** lunch at noon.*
AVOID The jury had **its** lunch at noon.

*If desired, the pronoun may be omitted: The jury had lunch at noon.

The plural "their" is used with the collective noun "jury" because they are eating separate lunches and are not acting as a unit

BUT

USE	The jury announced **its** verdict.†
AVOID	The jury announced **their** verdict.

When a jury arrives at a verdict, it is acting as *one*, or as a unit. Therefore the singular pronoun "its" is used as the noun substitute for the collective noun "jury."

When a collective noun is made plural (juries, companies, classes, teams, etc.), a plural pronoun would always be used as a substitute.

The **juries** announced **their** verdicts simultaneously in New York and in Chicago.

Recognizing collective nouns not only enables you to choose pronouns more accurately but also helps in Section Six when you're asked to be sure that nouns and verbs are in "agreement."

REPLAY UNIT 28

List collective nouns in the seven blanks below. See how many you can think of without looking back to the preceding pages.

1. ____________ 5. ____________
2. ____________ 6. ____________
3. ____________ 7. ____________
4. ____________

8. A pronoun that substitutes for a collective noun is usually (a) singular (b) plural
9. When the members of the group named by a collective noun are acting as separate individuals, the pronoun is (a) singular (b) plural
10. When the group named by a collective noun is acting as a unit, a pronoun that substitutes for it is (a) singular (b) plural

Underline the correct pronoun in parentheses.

11. Aerojet Corp. has (its/their) offices on the third floor.
12. The Accounting Department is on the third floor; (they/it) will be open until five o'clock.
13. Anderson, Temple, Inc., insists on having (its/their) bills paid promptly.

†If desired, the pronoun may be replaced with an article: The jury announced *the* verdict.

14. The Internal Revenue Service revised (its/their) forms again.
15. Her family are taking (its/their) vacations at different times this year.
16. The Board of Directors will hold (its/it's/their) annual meeting on Wednesday, May 7.
17. The city should regulate (its/it's/their) hiring policies more carefully.
18. Her family is taking (its/their) vacation in August this year.
19. The committee will review the new data at (its/it's/their) next meeting.
20. This firm is old, but (its/it's/their) management team is progressive.

See if your answers match those on page 316.

READ UNIT 29

A WHODUNIT

All of you are familiar with the whodunits of the movies and TV. Those of you who are mystery fans have read many of them. The best known of the English-language whodunits is the Mystery of Who and Whom. (Although it qualifies as a mystery, it's not quite a thriller.) Often those who try to use *whom* correctly in speech use it where *who* belongs, and others mumble so that listeners will not be sure whether they heard *who* or *whom*.

Many well-educated people still use *who* and *whom* traditionally, but the pendulum is swinging toward *who*. A few language authorities recommend that students no longer be asked to solve the Who/Whom Mystery and that *whom* be eliminated. A larger number of experts, however, suggest that *whom* be eliminated just from speech. They agree that when writing on a formal level, the traditional distinction between *who* and *whom* should be made. This middle-of-the-road approach is desirable for business communications. For most written communications, therefore, take a few seconds to determine whether *who* or *whom* should be used. In that way, you'll add to your confidence about the correctness of your writing. In speech, avoid *whom* unless you're certain it's the correct choice.

You can be right nearly all the time when deciding between *who* and *whom* if you use either of these two methods:

METHOD 1 Use **who** for the subject of a verb and **whom** for the object of a verb or of a preposition. The subject tells who or what DID the action referred to in the verb. The object of the verb tells who or what RECEIVED the action. The object of a preposition completes a prepositional phrase.

(a) *Who* gave me the portfolio?
(*Who* is the subject of the verb *gave*.)

(b) He is the man *who* gave me the portfolio.
(*Who* is still the subject of the verb *gave* even though the sentence is now longer.)

(c) He is the man *who* I think gave me the portfolio.
(*Who* is still the subject of the verb *gave* even though a two-word clause interrupts the subject and verb.)

(d) To *whom* did he give the portfolio?
(*Whom* is the object of the preposition *to*).

(e) He is the man *whom* you should marry.
(*You* is the subject of the verb *should marry—you* tells who is doing the action; *whom*, the receiver of the action, is the object of the verb *should marry*.)

METHOD 2 Change the order of the words so that you can replace the questionable **who** or **whom** with **he** or **him**. If **he** sounds right, use **who**; if **him** sounds right, use **whom**.

(a) He is the man (who/whom) gave me the portfolio.
(*He* gave me the portfolio—not *him* gave me the portfolio; therefore, choose *who*.)

(b) Ed Wahl, (who/whom) we understand visited your office yesterday, is an expert on data processing.
(Ed Wahl is an expert on data processing; we understand *he* visited your office yesterday—not *him* visited; therefore, choose *who*.)

(c) Jack Washington, (who/whom) Prudential hopes to hire, was trained at Harbor College.
(Jack Washington was trained at Harbor College; Prudential hopes to hire *him*—not *hire he*; therefore choose *whom*.)

(d) I'll assign the report to (whoever/whomever) I wish.
(I'll assign the report to *him*; therefore, choose *whom*.)

(e) He is the one (who/whom) should do the work.
(*He* should do the work; therefore, choose *who*.)

Use whichever of the two methods seems easier, or try both; you should get the same answer, of course. By the time you finish the practice sentences on the next page, you'll feel like an expert. No matter which method you are applying, the following hints are also useful.

- When the sentence is a question, try to imagine it as a statement before making the who/whom choice.

Who/Whom are you going with?

(Change to statement—You are going with who/whom. *Him*, not *he*, sounds right; and since "with" is a preposition, the object *whom* must be used.)

- When a sentence is worded so that *whom* would be the object of a verb or of a preposition but *who* would be a subject to a verb, choose *who*.

You should go with (whoever/whomever) is ready first.
Him sounds right after *with; whom* would be the object of the preposition.

BUT

He sounds right before *is; who* is needed as the subject of the verb.

THEREFORE

Use *who*—You should go with *whoever* is ready first.
Allow the subject to take priority over the object.

As my assistant, I want *whoever* gets here first.

Use *whoever* as the subject of "gets" (rather than *whomever* as the object of "want").

REPLAY UNIT 29

Insert *who* or *whom* in the blanks. For maximum benefit from this Replay, apply one or both of the two methods for choosing *who* or *whom*. **Practice**, not guessing, makes perfect.

1. ______ would you like to accompany me?
2. The man ______ I met at your office is an engineer.
3. ______ do you prefer for the job?
4. Give the scholarship to the one ______ needs it most.
5. They are the ones ______ are needed for the job.
6. He is the boy ______ I took to be your brother.
7. Each candidate should support ______ ever the convention chooses.
8. I don't know ______ will do the work.
9. ______ ever is willing to do the work will be given the responsibility.
10. ______ would you like to go with you?
11. I am the one ______ must make the decisions.
12. Give it to ______ ever gets there first.
13. My secretary is the one ______ I believe was chosen.
14. Give the scholarship to ______ ever needs it most.
15. The secretary ______ you sent for is a very nervous man.
16. Ms. Cates is the accountant ______ I feel would be best able to devise a new system.

17. Mrs. Siegel, ________ we met for the first time last week, is a talented artist.
18. Eric Smith is the one ________ helped me most.
19. Mr. Berman, ________ I told you about last week, will speak on the subject of laser beams.
20. We selected James Madden, ________ we know has been active in many organizations.
21. ________ do you suppose will get the position?
22. ________ do you think we should get to make the investigation?
23. The question of ________ should handle the advertising will be discussed.
24. Give the package to ________ ever can identify it.
25. We'll give a trip for two to Hawaii to ________ ever makes the most sales.

See if your answers match those on page 316.

READ UNIT 30

EVERYBODY NEEDS MILK

This title was a California radio and TV commercial a few years ago. Was it *every body* or *everybody*? That's just one type of question occurring with the use of indefinite pronouns—pronouns that don't refer to a definite person or thing.

each	anybody	somebody	nobody	few
everyone	anyone	something	all	many
everybody	anything	none	any	others
everything	someone	no one	both	several

ONE WORD OR TWO

Except for *no one*, the compound indefinite pronouns listed above are one word. In some sentences, however, these expressions are not pronouns at all; instead, the first part is an adjective and the second, a noun. Then two words are used. Fortunately it isn't necessary to analyze the grammar to determine whether to use one word or two. Just apply these two easy principles:

(a) *Any one* and *every one* are two words when the word **of** follows.

1. *Any one of* you may attend the conference.
2. *Every one of* you may attend the conference.

BUT

3. *Everyone* may attend the conference.

(b) Any of the compound expressions listed above are two words when your own good judgment indicates a special meaning.

1. Dr. Cutup took the medical students to the morgue and told them to examine *any body*.
2. The man I met on Muscle Beach has *some body*.
3. Phyllis Diller said, "(Every body/Everybody) needs milk."

Both are correct since *every body* stresses the needs of the body for milk, and *everybody* stresses that all people need milk. The double meaning is probably why the sentence was selected by the dairy industry for use in its commercial.

THE SINGLES SCENE

Although the following words are frequently indefinite pronouns, when the first four are used just before a noun, they are adjectives. Fortunately again, it isn't necessary to identify the part of speech—just remember that they are all **SINGULAR**.

each, every, either, neither (pronoun or adjective), everyone, everybody, nobody, somebody, someone, anyone, anybody, one, no one

Even though "everyone" might include five hundred people, it is still singular because it refers to each one acting individually. Therefore, when another pronoun substitutes for one of the words listed above, that pronoun must also be singular. (In informal speech this rule is frequently violated, but it is still considered important in formal writing.)

USE	Everybody went to his seat.
AVOID	Everybody went to their seat.
USE	Every worker should use his own tools.
AVOID	Every worker should use their own tools.
USE	Each of the rooms has its own heating unit.
AVOID	Each of the rooms has their own heating unit.

Since *everybody, every,* and *each* are singular, the singular pronoun *his* or *its* is used instead of the plural *their*.

GENDER IN MODERN BUSINESS COMMUNICATION

Gender refers to male, female, and neutral words. You may have noticed that in the two sample sentences above, *his* is used even though there may have been some females who went to their seats or who had tools. The English language, rich as it is, makes no provision for a special neutral gender third person singular pronoun (except for *it*).

When both a masculine and a feminine pronoun would be logical, choose from these alternatives:

(a) Use *his, he,* or *him* to represent both sexes. Masculine pronouns have long been accepted as carrying both meanings; that is, male only, as well as male and female.

Each person did *his* work quietly.

(b) Use *"his or her," "he or she,"* or *"him or her."* Using both pronouns is grammatically correct but sometimes sounds awkward.

Each person did *his or her* work quietly.

(c) Reword the sentence so that a pronoun isn't needed, provided you don't change the meaning or tone intended.

Each person did the work quietly.
Each person worked quietly.
Everyone worked quietly.

(d) Another good solution occurs when it's possible to reword the sentence so that the plural may be used correctly.

The people went about **their** work quietly. (Since "people" is plural, the plural pronoun "their" is acceptable.)

BUT

Don't compromise by using the nonstandard form:

Each person did *their* work quietly. NO!
Everyone did *their* work quietly. NO!

AND

To avoid the risk of offending a client, customer, or co-worker, develop sensitivity to "sexism" in language.

REPLAY UNIT 30

Circle the correct form in the following sentences.

1. (No one/Noone/No-one) but the owner knows the combination to the safe.
2. Please distribute the flyers to (any one/anyone) who wants them.
3. (Every body/Everybody) should learn how to type.
4. He asked that (some one/someone) from this office visit his store.
5. (Any one/Anyone) of you is qualified to prepare the report.
6. Although it was assumed that the pilot was killed, (no body/nobody) could be found in or near the wreckage.
7. Each person did (his/their) work quietly.
8. Everyone should complete (his/their) assignment before Friday.
9. Every woman in this office needs (their, her) own phone.

10. If (some one/someone) needs more paper, (he/they) should fill out Form 26.

11. Whenever (some body/somebody) loses (his/their) keys, (he/they) should come in through my private office.

12. (Every one/Everyone) of the members must vote.

13. (Every one/Everyone) should write (their/his or her) name in the upper right corner.

14. Each senator must use (his/their) own funds for this project.

15. (Every body/Everybody) must make (his or her/their) own decision.

In the blank on the left, write the letter of the form most suitable for modern business communication. In the next blank, write the letter of the only form that is **incorrect.**

Best	Wrong	
c	b	16. (a) A person can usually improve if he really tries. (b) A person can usually improve if they really try. (c) People can usually improve if they really try. (d) A person can usually improve if he or she really tries.
A	D	17. (a) When customers express their dissatisfaction, listen courteously. (b) When a customer expresses his dissatisfaction, listen courteously. (c) When a customer expresses his or her dissatisfaction, listen courteously. (d) When a customer expresses their dissatisfaction, listen courteously.
c	b	18. (a) Every one of the contractors submitted his or her bid today. (b) Every one of the contractors submitted their bids today. (c) The contractors all submitted their bids today.
B	c	19. (a) Did anyone here lose his notebook? (b) Did any one of you lose your notebook? (c) Did anyone here lose their notebook?
C	B	20. (a) Everyone should write his name on the form immediately. (b) Everyone should write their name on the form immediately. (c) Everyone should write his or her name on the form immediately.

Check answers on page 317.
The Pop Quiz for Units 28–30 is on page 271.

Checkpoint

Handy Pronoun Display

	Singular			*Plural*		
Person	***Subject***	***Object***	***Possessive***	***Subject***	***Object***	***Possessive***
1st	I	me	my, mine	we	us	our, ours
2nd	you	you	your, yours	you	you	your, yours
3rd	he, she, it	him, her, it	his, hers, its	they	them	their, theirs
Other	who whoever	whom whomever	whose	who whoever	whom whomever	whose

Along with our rapidly changing technologies and lifestyles, language styles also change. As with all kinds of changes, there are those who resist and use their energies defending the old; others are at the head of the parade. When selecting language style for business or profession, being in the middle is advisable. By being "in the middle," you have confidence that readers and listeners can concentrate on your message and not be distracted by something unusual in your language style.

When preparing a report for the president of your company or a letter to an important customer, you should not have to ponder over whether to use *I* or *me*, *who* or *whom*, and so on. You should **know** how to decide. When with your business and professional colleagues, you should not have to be self-conscious about your grammar. You should be **confident** that your grammar is correct.

Pronouns are often a trouble spot for people who are comfortable with Standard English **and** for those who grew up using a dialect or another language. When these principles of pronouns are mastered, you will be using pronouns in a manner acceptable to most successful business people of this decade.

Place a check in the blank when you understand the principle.

______ 1. Use subject pronouns as subject of a verb and object pronouns as object of a verb or of a preposition. Devices are included in this section that make it easy to apply this principle. (Units 25, 26, 29)

______ 2. Use a "self" or "selves" pronoun only when another pronoun will not make sense in that position in the sentence. When you do use it, be sure a standard spelling or pronunciation is used. (Unit 26)

______ 3. Possessive pronouns that refer to specific people or animals never have apostrophes. Possessive pronouns that do not refer to specific people (those ending in *one* or *body*) do have apostrophes. Distinguish between possessive pronouns and contractions. Some of them sound alike but are spelled differently. (Unit 27)

______ 4. Pronouns should agree in number (singular or plural) and in gender with the noun or other pronoun for which they are substituting. (Units 28, 30)

______ 5. Compound indefinite pronouns are written as one word. Simple methods enable you to determine when the expression is not a pronoun and requires two words instead of one. (Unit 30)

Additional grammar explanations regarding pronouns appear in the Grammar Supplement in the back of this book.

Special Assignment

A British steamship company that operated vacation cruises wrote to the British Admiralty asking permission for its officers to wear swords, as did British naval officers, feeling that it would add a touch of "class." The Admiralty did not like the idea but didn't want to refuse permission. The reply from the Admiral's office, therefore, granted permission, adding, "provided the swords are worn on the right side." The cruise company dropped the idea because they felt foolish about writing again to ask for clarification of the expression "the right side." This was probably an example of intentional lack of clearness, since the Admiral's office achieved its purpose.

Perhaps this reminds you of when you were a child and an elderly relative asked, "Honey, do you know you put your left shoe on your right foot?" After you looked at your feet in confusion, the double meaning of "right foot" was explained and everyone guffawed.

These little stories are to remind you of the importance of clearness in writing—especially business writing. Concentrate on writing clearly as well as correctly when completing this brief report.

Choose a type of work that interests you: fashion merchandising, secretarial work, business management, data processing, law, teaching, auto mechanics, or whatever. Interview a worker in your chosen field. Prepare questions in advance, such as: What does the work consist of? What is a typical daily routine? What about salary? Advancement possibilities? What are the advantages and disadvantages to this kind of work? What kind of training do you need to qualify? Include only the ideas that interest you. What do *you* want to know about this kind of work? Summarize the information you obtain and prepare a typewritten report of 175 to 200 words. Start with a topic sentence that tells the reader in an interesting way what the report is about. Here are sample topic sentences:

1. "Accounting is an excellent career for women," states my friend, Mary Lopez, who works in the controller's office of Hughes Aircraft Company.
2. I learned a great deal about the work of an accountant through interviewing Mary Lopez, who works at Hughes Aircraft Corporation.
3. Like other careers, teaching has many advantages as well as disadvantages.

Follow through with what you promised in your topic sentence. For example—The rest of the report for No. 1 would give reasons why accounting is a good career for women. With the No. 2 opening, the writer would describe the work of an accountant. The report beginning with No. 3 should inform the reader of the advantages and disadvantages of being a teacher.

Stay on the track right to the end. Here are sample closing sentences. The closing should be a conclusion or a summary of what went before.

1. Business management sounds like an exciting career filled with challenges and rewards.
2. Because of the long hours and the problems of dealing with difficult customers, I've decided to change my career goal; I no longer want to become a retail store manager.
3. Now that I've learned about the training for the profession of neurosurgery, I realize that I do not have the tremendous drive, ambition, and ability required to prepare for this career.

Submit this report to your instructor on ____________. (date)

Proofreading for Careers

A memo, also called an interoffice memorandum, is a message to one or more members of the same organization. Most organizations have specially printed forms for memorandums. Below is a memo typed on a typical interoffice memorandum form. Please show the necessary corrections of the grammar, spelling, sentence construction, capitalization, and typing.

NEW AGE COMPUTER CORPORATION

DATE: June 30, 198__
TO: New Employees
FROM: Ruth Warren, Personnel Director
SUBJECT: New Age Employment Policies

Welcome to the New Age Computer corporation. We hope you will find your employment here enjoyable and personaly rewarding. These policies will help you understand our operation, and it will help you to realize the importance of you're job.

- **Security** Each new employee should promptly obtain their permanent identification card from the personnel department. A Guard is on duty around the clock. Every one who enters the building is required to show their identification card.
- **Absenteeism** Employees are expected to be in their assigned departments and ready to begin work at 8:30. If you are unable to come to work, please call my Administrative Assistant, Mr. Harrison, at Ext. 711, or myself. Him or I will inform the department Manager to who you have been asigned.
- **Loyalty** Loyalty is expected from all employees, information about new products or financial matters should not be disclosed to the publick. Any one who passes on such information is subject to immediate dismissal.
- **Smoking** Because most New Age people do not smoke, smoking is not permitted in the cafeteria, employees may, however, smoke on the terace outside the cafeteria.

If you have any questions about these policies, please see Mr. Harrison or myself.

What's your proofreading score?
Excellent __ Good __ Fair __ Undistinguished __

AMERICAN GRAFFITI
(In a Washington Phone Booth)

Who should I call first?
An English teacher!

Practice Quiz

Circle the correct pronoun.

1. Her family is taking (its/it's their) vacation in August this year. (Unit 28)
2. Her family are taking (it/it's/their) vacations at different times this year. (Unit 28)
3. The Internal Revenue Service revised (its/it's/their) forms again this year. (Unit 28)
4. The Data Processing Department is on the third floor; (it is/they are) open every weekday. (Unit 28)
5. (No one/Noone) regrets this incident more than (I/me/myself). (Units 26 and 30)
6. Is Miss Jones better qualified than (her/she)? (Unit 26)
7. Johnson completed more of the programming than (I/me). (Unit 26)
8. No group is better liked than (they/them). (Unit 26)
9. The property on Adams Avenue was recently purchased by Goldman, Martinez, and (me/I/myself). (Unit 25)
10. Neither Mr. O'Connor nor (I/me/myself) will visit the Akron plant this year. (Units 25 and 26)
11. (Some one/Someone) from your office left (his or her/their) keys at the reception desk. (Unit 30)
12. (Every body/Everybody) in this office is being asked to do (his/their/they're) own thinking. (Unit 30)
13. The new machine on the counter is (your's/yours). (Unit 27)
14. If you prefer (hers/her's/hers'), you could request an exchange. (Unit 27)
15. (Who/Whom) did you say will handle the new account? (Unit 29)

16. The new member was seated between Johnson and (me/I/myself). (Units 25 and 26)

17. (Who/Whom) was he talking to when he disclosed our plans? (Unit 29)

18. (Him/He) and Kahn will discuss it with (whoever/whomever) you suggest. (Unit 29)

19. General Jones questioned the colonel and (I/me). (Unit 25)

20. If (anyone/any one) of the boys would like a ticket, (he/they) may have one. (Unit 30)

21. He (hisself/himself) thinks (it's/its) too late. (Unit 26)

22. That young man is one upon (who/whom) you can rely. (Unit 29)

23. We sent letters to all (who/whom) we thought might visit our showroom. (Unit 29)

24. The manager offered a prize to (whoever/whomever) sold the most policies. (Unit 26)

25. Those blueprints are (their's/theirs/theirs'). (Unit 27)

SECTION 6

LOOKING FOR THE ACTION? THEN STUDY VERBS!

After Completing This Section, You Will:

- *Use verbs so that the time of the action or existence is expressed correctly.*
- *Use subjects and verbs that agree in number and person.*

Every sentence has at least one verb. The verb tells either—

(a) what the subject DOES or HAS (action verb)

or

(b) what the subject IS (existence verb).

Verbs are the action words and the existence words of sentences. Some verbs consist of one word, as in these sentences:

(a) The personnel manager *interviewed* two applicants. (action verb)
(b) The personnel manager *was* in the office. (existence verb, also known as state-of-being verb or linking verb)

Other verbs consist of two or more words—a main verb and one or more helping verbs—as in these sentences:

(a) The personnel manager *has been interviewing* those applicants all day. (Action verb *interviewing* is main verb; *has* and *been* are helping verbs.)
(b) The personnel manager *should have been* in the office today. (Existence verb *been* is main verb; *should* and *have* are helping verbs.)

To use verbs correctly, these major principles must be learned:

1. Every verb has a basic form called the **infinitive**. The infinitive consists of the word "to" plus the spelling of the verb that is most easily located in the dictionary as an entry word. Examples are—**to eat, to dance, to have, to be, to love, to work, to type**. The basic verb form that follows the word "to" is what to look for in the dictionary when you want to find other forms of that verb.
2. The verb form you use should be based on:
   - whether the *time* of the action or existence is
     now
     before
     later
   - whether the subject of the verb is
     singular or
     plural

- whether the subject of the verb is

1st person	I or we
2nd person	you
3rd person	he, she, it, or they
other pronoun	such as *who, everybody*, or *which*
any noun	such as *story, supervisor*, or *pen*

Only the important rules governing verb forms are included in this section. If, however, you grew up in a community where Standard English is used in everyday conversation, you will find that some of these rules do not require your study because you automatically use the correct form.

BUT

If you grew up in

(a) a non-English-speaking country,
(b) a community where English is a "second language" for many of the residents, or
(c) a community where regional or ethnic English is usually used,

THEN

Getting into the habit of applying these rules might make the difference between success and failure in your career. Most of the nonstandard verb forms brought to your attention are the ones that are very noticeable in a business environment. Therefore, it is especially important to apply the rules studied in this chapter to your everyday speech so that you'll automatically use the standard forms.

> "For there be women, fair as she,
> Whose verbs and nouns do more agree."
> —*Bret Harte*

READ UNIT 31

TIMELY TIPS

Action or existence expressed by verbs takes place in three principal time periods: past, present and future. (These are called "tense" in traditional grammar books.)

Some action or existence, however, includes more than one period of time. For example, something may start in the past and continue into the present. Other action might start in the present and continue into the future. Our verbs have enough variety in forms so that "time" is communicated by the way the verb is expressed. Here are just some of the ways in which time is expressed. The *time* of the action is interpreted below each sentence.

I *have been working* on this for days.
(I started working in the past and am still working now in the present.)

I *worked* on this for days.
(I started in the past but am not working on it now in the present.)

I'll work on this for days.
(I will continue to do the work in the future.)

Most grammar rules for expressing time of verbs do not need to be included in this section because adults usually select the appropriate verb form without having to think about it. You are asked to study, therefore, only those rules that result in the elimination of errors that **do** occur in students' work.

VERB CATEGORIES: REGULAR AND IRREGULAR

A *regular* verb adds *s, ed,* or *ing* to the basic (infinitive*) form in the usual (or regular) way:

BASIC FORM	WALK
Add *s*	walks
Add *ed*	walked
Add *ing*	walking

An *irregular* verb does not follow any set pattern:

BASIC FORM	BEGIN
Add *s*	begins
Don't add *ed*; instead change to	began or begun
Double last letter before adding *ing*	beginning

It is important to understand the system in your dictionary that reveals whether or not a verb is regular. Look up the verb **walk** right now. Be sure that you look at the VERB *walk,* not the noun.

In some dictionaries, you will find *walked, walking, walks* right next to the entry word. This means that all the forms of **regular** verbs *are* included in the dictionary.

In other dictionaries, you will not find the three other forms of *walk.* This means that the changed forms of regular verbs are not included in those dictionaries. When you cannot find the other forms of a verb next to the entry word, it is regular and you follow the same pattern as you would for *walk.*

For **irregular** verbs, every good dictionary includes the various forms. Please look up **begin** in your dictionary now.

RULES FOR REGULARS

1. TIME: PAST
   If the action was in the past or if **has**, **have**, or **had** precedes the verb, add **ed** to the basic verb form.

   He *walked* home this morning.
   He *has walked* home every morning this week.

*See page 299, No. 1.

2. TIME: IN PROGRESS

If the action is or was in progress, use a helping verb (**is, am, are, was, were, have been, has been, had been**) before a main verb that ends with **ing**.

They *are walking* home today.
He *was walking* home when the accident occurred.
You *have been walking* home every day for a year now.

3. TIME: FUTURE

When **will, shall, would**, or **should** precedes a verb, use the basic verb form.

He *should walk* home today.
He *will walk* home today.
I *would walk* home today if I could.

4. TIME: PRESENT

When there is no helping verb, use the basic verb form plus an **s** if the subject is:

(a) any singular pronoun except **I** or **you**.

He walks home every day.
Somebody walks home every day.

(b) any singular noun.

The *man walks* home every day.
My *patient seems* better now.

When there is no helping verb, use the basic verb form (**no s**) if the subject is:

(a) **I** or **you.**

I walk home.
You walk home.

(b) any plural pronoun.

We walk home.
They walk home.

(c) any plural noun.

The *men walk* home.
The *attorneys seem* happy.

REPLAY UNIT 31

Write in the blank a correct form of the verb in parentheses. Notice the **time** of the action indicated in parentheses at the end of the sentence. Some of the sentences require a helping verb.

Sample They (talk) ____talked____ every day on the telephone last week. (past)

1. He (type) ________ very accurately. (present)
2. The vice president (need) ________ to make a decision at once. (present)
3. The merchandise (arrive) ________ yesterday. (past)
4. They (sail) ________ to Catalina this week. (*ing* form)
5. The floor is slippery because the custodian (wax) ________ it today. (past)
6. Mr. Martinez (paint) ________ beautiful landscapes. (present)
7. That secretary (consult) ________ the dictionary frequently. (present)
8. The foreman (climb) ________ up the ladder in order to reach the carton. (past)
9. We (gain) ________ some insight into why the problems exist. (past)
10. They (want) ________ something they cannot have. (present)
11. She (want) ________ something she cannot have. (present)
12. He (want) ________ something he could not have. (past)
13. She (want) ________ something she cannot have. (future)
14. They (offer) ________ the employees a raise. (past)
15. He believes he (select) ________ a business career. (future)
16. Everyone (look) ________ tired today. (present)
17. Everyone (look) ________ tired yesterday also. (past)
18. The manager (consider) ________ that problem tomorrow. (future)
19. The manager (consider) ________ that problem now. (*ing* form)
20. You (want) ________ those items yesterday. (past)
21. I (stay) ________ at that hotel every July. (present)
22. They (stay) ________ there now. (*ing* form)
23. We (watch) ________ television all day. (past)
24. You (need) ________ these items in order to build the stereo. (present)
25. Marketing people (talk) ________ about local, regional, and national markets. (past)

Check your answers on page 317.

READ UNIT 32

DELINQUENT VERBS

Delinquents—verbs that won't behave as they should—are introduced in this unit. The best way to deal with these irregular verbs is to:

(a) Recognize that they are irregular.
(b) Look them up in the dictionary when uncertain of form or spelling. To obtain information about irregular verbs, look up the "basic" form, which is the one in Column A.

Column I		***Column II***	***Column III***
A	***B***	***Simple Past***	***Past Participle***
begin	begins	began	begun
break	breaks	broke	broken
choose	chooses	chose	chosen
do	does	did	done
drink	drinks	drank	drunk
eat	eats	ate	eaten
fly	flies	flew	flown
give	gives	gave	given
go	goes	went	gone
ring	rings	rang	rung
rise	rises	rose	risen
run	runs	ran	run
see	sees	saw	seen
speak	speaks	spoke	spoken
take	takes	took	taken
wear	wears	wore	worn

INSTRUCTIONS FOR COLUMN I

Side A The words under **A** are the basic forms. They are used with the helping verbs *will, shall, would, should, do, did, does, might, may, can,* and *could,* no matter what the subject is.

He did *begin.* I *should begin.* She *will begin.*

The **A** words are also used without helping verbs when the subject of the verb is *I, you,* any plural pronoun, or any plural noun.

I *speak* the truth. Those glasses *break* easily.
The two little boys *break* everything.

Side B The **B** words (which end in *s*) are used with any **singular** subject except *I* or *you.*

Everybody *begins* at the same time. She *begins* at 9 a.m.
Mr. Mosk *begins* later.

INSTRUCTIONS FOR COLUMN II

Use these words for expressing the past **without a helping verb**. Use them with all subjects BUT NEVER WITH HELPING VERBS.

She *began* the work. The glass *broke.* They *broke* the glass.

INSTRUCTIONS FOR COLUMN III

Use these forms **with a helping verb**, such as *have, has, had, been, was, were.*

I *have begun* the work. The glass *was broken.*
The best fabrics *will have been chosen* by that time.

> The verbs in these three columns are among the most commonly misused verbs. Since other irregular verbs are used incorrectly also, careful writers refer to the dictionary to make sure of the Standard form.

SENTENCES WITH TWO OR MORE VERBS

When a sentence has two or more verbs, we generally express them in the same "time" (or tense).

Present She **thinks** that I **am** a millionaire.

Past Mr. Hanson **wrote** me a note in which he **implied** that I **passed** the accounting test.

In the sentences above, the same time is used for verbs in the same sentence.

However, to express a general truth or something that is still happening now, use the present, even if the verb in another part of the sentence is in the past.

Right Ms. Cortese **told** us that Tokyo **is** larger than New York City.

Wrong Ms. Cortese **told** us that Tokyo **was** larger than New York City.
(Even though she told us in the past, Tokyo is still larger.)

Right The sales manager **had demonstrated** several times that our equipment **performs** better than any other on the market.

Wrong The sales manager **had demonstrated** several times that our equipment **had performed** better than any other on the market.
(If our equipment **had performed** better, then perhaps it doesn't perform very well now in the present.)

Right What **are** the titles of the books you **borrowed** from the library?

Wrong What **were** the titles of the books you **borrowed** from the library?
(Although you borrowed the books in the past, the titles are still the same.)

REPLAY UNIT 32

Refer to the three columns on page 138 when necessary in order to write the correct form of the verb in the blank. Change only those verbs that are incorrect. Do not make a change just because you feel another word would sound better. If all the verbs in the sentence are correct, write C in the blank.

1. I thought it was broke, but it had just worn out. ______
2. He had just began work when the bell rang. ______
3. He choose the best one for himself. (present) ______
4. He had chose the best one for himself yesterday. ______
5. She do all the complaining. (present) ______
6. He has drunk all the tea. ______
7. Everyone in this office eat donuts. (present) ______
8. Smith done a good job of figuring those prices. ______
9. They flies to New York every year. (present) ______
10. He give his secretary an increase in salary. (past) ______
11. They had went to the convention before they saw him. ______
12. We have rose up together to fight the enemy. ______
13. The treasurer has ran away again. ______
14. She had wore the new suit to the office again. ______
15. When he speak, everyone listens. ______
16. I seen him when he was reading the letter. ______
17. I have taken it with me. ______
18. The men always wears suits to the office. ______
19. She had saw that file before. ______
20. I spoke to him on the telephone yesterday. ______
21. They ate lunch on the plane. ______
22. The maid rung the dinner bell 15 minutes ago. ______
23. The maid has already rung the dinner bell three times. ______
24. The puppy ran away again. ______
25. They always gives the assembler three days to complete that kind of job. ______
26. The officer said that obeying traffic laws was necessary for accident prevention. ______
27. Seth knew that New York was the biggest city in the United States. ______
28. All the students agreed that the new courses were more difficult than the old ones. ______

29. What were the names of the authors of the books on the best-seller list in last Sunday's *Times*? ______

30. Robert Stern said the new Hewlett-Packard 150 was being used now in his office. ______

31. They were former employees of General Motors. ______

32. He taught us that rivers flowed into oceans. ______

33. We learned that no scientist knew with certainty about the origin of the universe. ______

34. Shirley explained that there were just two kinds of people in the world today. ______

35. The Chicago Police Department has still not discovered where the 5 million dollars was. ______

Please check your answers on page 317.

READ UNIT 33

DICTIONARY DATA

The verbs that don't behave as they should—the irregulars—included in Unit 32 are just some of over 100 verbs that change forms in nonregular ways. By using English over a period of years, you've automatically mastered most of the irregularities. However, if you're like most people, a few still cause you to hesitate. There are probably also a few you use incorrectly without being aware of it.

Since there are more irregular verbs than would be practical to study, it is important to know how to interpret verb information given in the dictionary. When looking up a verb, you can find its "principal parts." The *principal parts* of any verb are:

1. Basic form—break, see, walk, and so on.
2. Simple past form—broke, saw, walked.
3. Past participle form—broken, seen, walked.
4. Present participle form—breaking, seeing, walking.

The explanations below will enable you to use the dictionary intelligently.

INFINITIVE

The basic form is easily recognized because it is the dictionary entry word for that verb. Samples are in Column 1, Side A, Read Unit 32.

PRESENT PARTICIPLE

The present participle is easy to recognize because it always ends with **ing**. Be aware, however, of spelling variations.

- Usually the *ing* is simply added to the unchanged infinitive form—see, seeing.
- Sometimes a final *e* is omitted before adding *ing*—love, loving.
- Sometimes a final letter is doubled before adding *ing*—win, winning.

SIMPLE PAST AND PAST PARTICIPLE

REGULAR VERBS

When a verb is regular, the simple past and the past participle are the same: just add *ed*—walk*ed* is both simple past and past participle. Therefore, we used *walked* or the *ed* form of any regular verb either with or without a helping verb. Both *I WALKED* and *I have WALKED* would be correct, depending on which sounds right to you to express a particular idea.

IRREGULAR VERBS

The explanation that follows REQUIRES THE USE OF YOUR DICTIONARY.

1. All the Column II verb forms in Unit 32 are simple past, which means they are used without a helping verb. In the dictionary, the form you see first after the entry word (sometimes after the *ing* form) is the simple past. This means it belongs in Column II and is used **without** a helping verb. Look up "begin" as an example and see where "began" appears in your dictionary.

2. The form in the dictionary **after** the simple past is the past participle and could be listed in Column III in Unit 32. Therefore, a helping verb **is required**. Notice how "begun" follows "began" in your dictionary.

3. If only one *past* form appears, then that form is both the simple past *and* the past participle. All you need to remember about that is to use the same form **with** or **without** a helping verb. Look up "bring." Notice that only one past form appears—"brought."

4. Sometimes you'll see the word *or* or *also* between two forms of the verb. This means either word is OK with or without a helping verb. Look up "broadcast." Notice that the simple past is either "broadcast" or "broadcasted." Since no other form follows (in most dictionaries), the same two words are also suitable for the past participle, as explained in Item 3 above.

5. Now look up "show." Notice that the first form after the entry (or after the *ing* word) is "showed," which means it's the simple past—the one used without a helping verb. Following "showed" is "shown or showed" in most dictionaries. Therefore, either of those two is used with a helping verb. With that particular word, if you use "showed," you can't go wrong; but if you use "shown" without a helping verb, you are in trouble.

6. Finally, look up "occur." You find the *r* is doubled before adding *ed* or *ing*. When you have the slightest doubt about whether to double the last letter before adding *ed* or *ing* or whether to drop a final *e* before adding *ing*, LOOK IT UP!

> **MEMO TO THE CONSCIENTIOUS BUT RESTLESS STUDENT**
> Did you take the necessary few minutes to look up each word suggested in the preceding six dictionary-use hints? Did you also refer to Columns II and III in Unit 32? If you did, then you now know how to use your dictionary in an intelligent manner to obtain information about verbs. It is not necessary to memorize hundreds or thousands of forms; just know how to look them up.

REPLAY UNIT 33

Fill in the blanks with the "principal parts" of these irregular verbs. Please use your dictionary whenever you have the slightest doubt about the correct form.

	INFINITIVE	PRESENT PARTICIPLE (*ing* ending)	SIMPLE PAST (no helping verb)	PAST PARTICIPLE (requires helping verb)
Sample	beat	beating	beat	beaten or beat
	1. bite	biting	bit	bitten
	2. blow	blowing	blew	blown
	3. burst	bursting	burst	burst
	4. come	coming	came	come
	5. cost	costing	cost	cost
	6. draw	drawing	drew	drown
	7. fall	Falling	Fell	Fallen
	8. forecast	FORECASTING	Forecast(ed)	Forecast(ed)
	9. forget	FORGETTING	FORGOT	FORGOTTEN
	10. freeze	Freezing	FROZE	FROZEN
	11. grind	Grinding	ground	ground
	12. hide	Hiding	Hid	Hidden
	13. lead	leading	led	led
	14. pay	Paying	Paid	Paid
	15. shake	Shaking	shoke	shaken
	16. sink	Sinking	sank	SUNK
	17. sing	singing	sang	SUNG

18. throw	throwing	thrww	thrown
19. win	winning	won	won
20. write	writing	wrote	written

Each of the sentences below has an incorrectly used irregular verb. Write the correct form in the blank. Use your dictionary to be sure of the correct spelling.

beatten 21. Roberta Simon has beat all previous sales records.

did 22. Ms. Jaffe done the job right.

Broken 23. Because Sam has broke one of the rules, the boiler burst.

hung 24. Laurie hanged the picture, but it fell.

Losing 25. For the past two years the company has been loseing money.

Please check your answers on page 317.
Then take the POP QUIZ on page 272.

READ UNIT 34

TO BE, OR NOT TO BE

"To be, or not to be: that is the question," is one of the most often quoted Shakespearean lines. The infinitive "to be" has yet another claim to fame: It is the most irregular verb in the English language—and probably the most commonly used. If you grew up in a community where nonstandard English is frequently spoken, give careful attention to the standard forms shown below. For those who customarily use Standard English in everyday neighborhood and family speech, just glancing at the forms will be sufficient.

The forms of the verb "to be" are **am, is, are, was, were, be, being**, and **been**. **Been** is always preceded by *have, had,* or *has.*

IF THE SUBJECT IS *I*

Use	*Avoid—As in These Sentences*	
I am going now.	I be	I be going now.
I was going home.	I is	I is going now.
I will (or shall) be going now.	I been	I been here before.
I have been here before.		
I will (or shall) have been gone long before he arrives.		

IF THE SUBJECT IS *YOU*

Use	Avoid—As in These Sentences	
Are you going now?	You be	You be going now?
You were my best friend.	You is	You is my best friend.
You will be going tomorrow.	You was	You was there yesterday.
You have been there before.	You been	You been to see him every day.
You had been gone for 15 minutes.		
You will have been to see him before I get there.		

IF THE SUBJECT IS ANY *SINGULAR NOUN* OR *SINGULAR PRONOUN* EXCEPT *YOU* OR *I*

Use	Avoid—As in These Sentences	
He is home now.	He be	He be home now.
She was home yesterday.	She be	She be coming over soon.
It will be there for a week.	It been	It been there for a week.
He has been gone for a week.	She were	She were here.
He had been gone all week when the letter arrived.		
Everybody is here.	Everybody be here. (unless it is a command)	
Everyone was going.	Everyone were going.	
Was anyone going?	Were anyone going?	
John is the new accountant	John be the new accountant.	
The book had been lost.	The book been on the table.	
The train was late.	The train were late.	

IF THE SUBJECT IS A *PLURAL NOUN* OR *PRONOUN*

Use	Avoid—As in These Sentences	
We (or They) are at home.	We be	We be going home.
We (or They) were going home.	We is	We is going home.
We (or They) will be here today.	They was	They was home.
We (or They) have been gone all day.	They been	They been gone all day.
We (or They) had been gone all week.		
We (or They) will have been here for a year when the new building opens.		
The boys have been late every day.	The boys has been late every day.	
The salespeople are early.	The salespeople is early.	
The books were lost.	The books was lost.	

Contractions of many of the expressions in this unit are appropriate for speech and for nearly all business writing.

We're shipping your order today.
or
We are shipping your order today. } Both forms are correct.

She's the new accountant.
or
She is the new accountant.
} Both forms are correct.

ARITHMETIC TIP
2 plus 2 *is* 4, but 2 and 2 *are* 4, and 2×2 *is* 4.
I don't know why; that's just the way it is . . . are?

REPLAY UNIT 34

If the form of the "to be" verb is correct, write *C* in the blank. Otherwise, make the necessary correction.

_____ 1. We was going to meet them in Miami for the meeting.

_____ 2. They be going to Miami for the meeting also.

_____ 3. How long you planning to be gone?

_____ 4. We been gone for a year and no one misses us.

_____ 5. We is pleased to announce a new policy.

_____ 6. We hope this new service be helpful to you.

_____ 7. We be enclosing your new catalog.

_____ 8. We're hoping to visit your new showroom soon.

_____ 9. I'll be happy to ship the toasters today.

_____ 10. You is required to pay for them within thirty days.

_____ 11. Was you present when the officers of the company entered the room?

_____ 12. Ms. Johnson were happy to receive the new calculator for her office.

_____ 13. Everyone in our office are planning to take the night flight to Dallas.

_____ 14. Do you think everybody is going?

_____ 15. The account books was misplaced because the clerk misfiled them.

_____ 16. The shipping clerks was late every day this week.

_____ 17. However, the salespeople is always on time.

_____ 18. The new saleswoman been meeting her quota every week.

_____ 19. I have been away from my office all day and missed your calls.

_____ 20. Was you away from your office yesterday also?

_______ 21. The women was waiting in the lobby.

_______ 22. Mark Allen be studying at Columbia University.

_______ 23. Mark has been at Columbia for three years.

_______ 24. Dr. Melanie Sweet be in Philadelphia today.

_______ 25. Nobody were there yesterday.

Answers are on page 318.

READ UNIT 35

THE SUBJECT-VERB COMBO

Every sentence has at least one subject-verb combination. The verb is the action or existence word, and the subject tells who or what is acting or existing. Every subject is either a noun or a pronoun. To apply the principles of good English usage reviewed in Read Units 36 and 37, you need to identify subject-verb combinations.

> I recommend these steps for locating the subject-verb combination:
>
> - Find the verb—Anita dances in the show. (*Dances* is the action word.)
> - Ask: "Who or what dances?" *Anita* dances.
> - Then *Anita* is the subject of the verb *dances*.

1. **The subject is usually at the beginning of a clause (a word group containing a subject and a verb).**

The student studies.	*Subject:*	student—tells *who* is doing the action.
	Verb:	studies—the *action*.

2. **When a sentence includes more than one clause, each clause has its own subject and verb.**

Some students study and others daydream	*Subjects:*	students, others
	Verbs:	study, daydream

3. **In questions, a helping verb precedes the subject; and the main verb follows the subject. The helping verb plus the main verb make up the complete verb.**

Do students study?	*Subject:*	students
	Verb:	do study

4. **Sometimes an introductory expression comes before the subject.**

In Schenectady, students do study.	*Subject:*	students
	Verb:	do study

5. **Describing words are sometimes used before the subject.**

Ambitious and diligent students study.	*Subject:* students *Verb:* study

6. **Sometimes a prepositional phrase separates the subject from the verb. The subject will never be in a prepositional phrase.**

A person *with many skills and talents* is fortunate.	*Subject:* person *Verb:* is *Preposition:* with *Prepositional phrase:* with many skills and talents

I suggest you take a moment NOW to review prepositions and prepositional phrases in Read Unit 5.

7. **When a sentence has more than one clause, one of the clauses may "interrupt" another; that is, one clause separates the subject and verb of another clause.**

A person *who has many skills and talents* is fortunate.	*Subject:* person *Verb:* is *Subject:* who *Verb:* has

Two Clauses: A person is fortunate/who has many skills and talents.

8. **Some subjects have two or more parts.**

Laziness and irresponsibility are not needed for success.	*Subject:* laziness, irresponsibility *Verb:* are needed

9. **A single subject may have two or more verbs.**

Robin sings and dances.	*Subject:* Robin *Verbs:* sings, dances

10. **Although the subject normally is before the verb, sometimes the order is reversed.**

In the supply cabinet will be found six typewriter ribbons and four stencils.	*Subject:* ribbons, stencils *Verb:* will be found
There are six ribbons in my desk drawer also.	*Subject:* ribbons *Verb:* are

11. **Although the subject is always a noun or a pronoun, it may look like a verb when it names an activity**. Remember that a subject tells who or what does the action or exists. Any word that tells who or what acts or exists is automatically a subject and therefore a noun or a pronoun.

Studying is my favorite pastime.	*Subject:* studying *Verb:* is

To run would be foolish.	*Subject:*	To run
	Verb:	would be

12. **Sometimes the subject is invisible; that is, it is understood but not expressed. The "understood" subject is always YOU and introduces a sentence that gives an order, or a command.**

Put a new ribbon on the typewriter. or Please put a new ribbon on the typewriter.	*Subject:* *Verb:*	You (understood) put

REPLAY UNIT 35

Underline the verb and circle the subject. When a verb is preceded by a helping verb, underline it along with the main verb. Remember that a sentence may have more than one subject-verb combination.

1. I enjoy the sunset.
2. The salespeople are doing their jobs well.
3. Do you read many magazines?
4. He did such a fine job that I could not find a mistake.
5. We voted for him and he was elected.
6. Under the circumstances we won't issue a statement now.
7. For the position of receptionist, a pleasing personality is important.
8. Advancement comes to workers who do their work well.
9. Hard, steady work turns daydreaming into reality.
10. Playing Monopoly is very enjoyable.
11. During the past five years, Burroughs sales have risen dramatically.
12. Success-oriented students study several hours a day.
13. Suresh is the one who presented the most accurate analysis.
14. Two older memory typewriters are in the rear of the word processing center.
15. Rachel Rothstein will probably receive the promotion because of her unusual mathematics ability.
16. Everyone in the shop is working on the blueprints.
17. A. B. Dick & Co. manufactures duplicating machines.
18. My assistant, who is on vacation this week, will gladly help you next week.
19. Would you like my secretary's help next week?

20. Considerable turnover in personnel is one of the problems prevalent in the fast-food business.

21. Both poverty and riches are a state of mind as well as of pocketbook.

22. The English Shoe Company distributes its shoes through over two thousand shoe stores all around the United States.

23. We are returning your money order and hope that you will take it to your local English Shoe store at 30711 Ganado Drive.

24. The typewriters, which have new ribbons, will be delivered to the new office at 2166 Clinton Avenue.

25. Chan & Lopez, Inc., has an office in the new building, but the name has not yet been placed in the lobby directory.

Check your answers on page 318.

READ UNIT 36

SUBJECT-VERB HARMONY—STEP 1

A subject and its verb should always be in harmony. This means they should get along with each other—or *agree*. To make a subject and verb agree, the first step is to locate the subject and decide whether it is singular or plural.

SINGULAR SUBJECTS

(a) A singular subject represents *one* person, place, thing, idea: *lady, book, computer, it, he, honesty, glass, this, that,* and so on.

(b) A singular subject may be a collective noun acting as a "unit" (Review Unit 28): *club, class, jury, herd, staff, name of a company,* and the like. A collective noun is like a single package that contains several items. If reference is made to the entire "package," the collective noun is singular.

(c) A singular subject may consist of two (or more) words joined by **or** or **nor** followed by a singular noun or pronoun: *vice president or general manager, Mary nor he, the boy or the girl, noun or pronoun.*

(c) A singular subject results when *each, every,* or *many a* precedes it. *Each accountant, every secretary, many a child.*

(e) A singular subject may be an indefinite singular pronoun: *each, every, one, either, neither, another,* or any pronoun ending with *one, body,* or *thing*—such as *anyone, somebody, everything.*

PLURAL SUBJECTS

(a) A plural subject represents more than one person, place, thing, idea: *ladies, books, computers, they, glasses, these, those,* and so on.

(b) A collective noun is a plural subject when the members act as separate individuals: *club, class, jury, herd, staff, family, faculty,* and the like.

The *faculty are* in disagreement about the new grading policy.

If reference is made to *part* of the "package" (or collective noun), the package is broken. Then you consider all the separate items in the package, making the collective noun plural. (Review Unit 28 now.)

(c) A subject is plural if it is made of two or more words joined by *and**: *man and woman, John and she, Remingtons and Royals.*

(d) A plural subject may consist of two (or more) words joined by **or** or **nor** followed by a plural noun or pronoun: *one boy or three girls, vice president or general managers, Mary nor they, nouns or pronouns.*

(e) *Both, many, several,* and *few* are plural.

SINGULAR OR PLURAL SUBJECTS

These words are singular or plural, depending on the concept they refer to: *all, none, any, more, most, some,* or a fraction such as *half.*

Some of it **is** on the table.
Some of them **are** on the table.

If the item referred to in the prepositional phrase can be counted, use a plural verb:

Half of the apples **are** on the floor.

BUT

If the item cannot be counted because it is in bulk condition, use a singular verb:

Half of the applesauce **is** on the floor.

REPLAY UNIT 36

In these sentences, the subjects and verbs are in harmony; that is, they agree in number (singular or plural). First underline the verb in each sentence; then circle the subject (which might be more than one word). In the blank, write *S* or *P* to show whether the subject is singular or plural.

1. Several groups were invited to the meeting. P
2. She wants all the papers now. S

*When *each, every,* or *many a* comes before the parts of a subject joined by *and,* the subject is singular:
Each boy and girl (singular subject)
Every woman and man (singular subject)

3. My family took separate vacations this year. ______

4. My family piled into the truck and left the old homestead. ______

5. Abel & Lopez, Inc., has an office in the new building. ______

6. Everyone was waiting in line quietly. ______

7. Everybody seems happy today. ______

8. The report and the letter were on my desk. ______

9. Either the report or the letter was on my desk. ______

10. Neither the principal nor his assistants are here now. ______

11. Both the secretary's story and the typist's story sound true. ______

12. Neither the secretary's story nor the typist's story sounds true. ______

13. The battery, the radio, and the antenna are missing. ______

14. In other words, everything is gone. ______

15. Neither Mr. Jones nor Ms. Goldberg will work on it. ______

16. Each person rides the elevator twice a day. ______

17. The class took their seats. ______

18. The jury arrived at a verdict in two hours. ______

19. Any one of us can do the job. ______

20. Both a lawyer and a layman should be present. ______

21. Half of the pie is gone. ______

22. Half of the children are absent. ______

23. All of the workers are on strike today. ______

24. All is quiet on the Western Front. ______

25. Most of the apples are gone. ______

See answers on page 318.

READ UNIT 37

SUBJECT-VERB HARMONY—STEP 2

In Read Unit 35 you reviewed how to locate the subject by first finding the verb and then asking who or what did the action. Then you practiced distinguishing between singular and plural subjects. The next step in subject-verb harmony is reviewing how to make the verb agree with its subject in number (singular or plural).

1. **Although most** *nouns* **are made plural by adding** *s,* **do** *not* **apply this principle to verbs. As you saw in Units 31 and 32, adding an** *s* **to a verb makes it suitable to use with a** *singular* **noun or a** *singular* **pronoun (except** *you* **or** *I*).

   If the subject is a singular noun or pronoun (except *you* and *I*), add *s* to the verb:

PRONOUNS	NOUNS
He run**s** five miles every day.	**Ms. Chandler** run**s** five miles every day.
This mean**s** a great deal to me.	This **honor** mean**s** a great deal to me.
Everyone arrive**s** at 9 a.m.	Either **Charles or Dolores** arrive**s** at 9 a.m.

   If the subject is a plural noun, plural pronoun, you, or I, do not add *s* to the verb:

PRONOUNS	NOUNS
We run five miles every day	Jim and Walt run five miles every day.
These mean a great deal to me.	The honors mean a great deal to me.
They arrive at 9 a.m. I arrive at 10 a.m.	The employees arrive at 9 a.m.

2. **When** *or* **or** *nor* **joins the elements of a subject, consider only the part of the subject closest to the verb.**

WHICH IS IT?	Neither Charles nor Dolores (arrives/arrive) at 9 a.m.
HOW TO FIND OUT?	Imagine eliminating all words before "Dolores." Then it's easy to decide: Dolores arrives at 9 a.m.
CORRECT FORM	Neither Charles nor Dolores **arrives** at 9 a.m.
WHICH IS IT?	Ms. Pitz, Mr. Schaefer, or two Kensington seniors (represent/represents) Buffalo in the annual contest.
HOW TO FIND OUT?	Imagine omitting all words before "seniors." Then it's easy to decide: seniors represent Buffalo in the annual contest.
CORRECT FORM	Ms. Pitz, Mr. Schaefer, or two Kensington seniors **represent** Buffalo in the annual contest.
WHICH IS IT?	Either Christine or I usually (sort/sorts) the mail at Kensington.
HOW TO FIND OUT?	Imagine starting with "I." Then see which form agrees with "I." I usually sort the mail at Kensington.
CORRECT FORM	Either Christine or I usually sort the mail at Kensington.

3. **Usually a group of words separating the subject from its verb does not affect the number (singular or plural) of the verb.**

Right The box of tools is on the table.
Wrong The box of tools are on the table.

The subject is the singular noun *box*. Therefore, the singular verb form (*is*) must be used. Ignore the prepositional phrase (*of tools*) when deciding on the number of the verb.

Right The supervisors, as well as the president, are here today.
Wrong The supervisors, as well as the president, is here today.

The subject is the plural noun *supervisors*. Therefore, the plural verb form (*are*) must be used. Ignore the prepositional phrase (*as well as the president*) when deciding on the verb form.

Right The reason for his difficulties seems clear.
Wrong The reason for his difficulties seem clear.

The subject is the singular noun *reason*. Therefore, the singular verb form (*seems*) is required. Ignore the interrupting prepositional phrase when deciding on the verb.

Right The reasons that his job was difficult seem clear.
Wrong The reasons that his job was difficult seems clear.

The subject is the plural noun *reasons*. That's why the plural verb form (no *s*) is required. Ignore the clause (*that his job was difficult*) separating *reasons* from *seem*.

4. **The words HERE and THERE are not subjects. In a sentence introduced by** *here* **or** *there*, **the subject usually follows the verb.**

Right There was a box of tools on the table.
Wrong There were a box of tools on the table.

The subject is the singular noun *box*. Therefore the singular verb form (*was*) is used. OF TOOLS is a prepositional phrase.

Right There were several boxes of tools on the table.
Wrong There was several boxes of tools on the table.

Because the subject is the plural noun *boxes*, the plural verb form (*were*) is required.

Right Here is the schedule.
Wrong Here are the schedule.

The subject is the singular noun *schedule*. Therefore, the singular form of the verb (*is*) must be used.

Right Here come my sisters.
Wrong Here comes my sisters.

The subject is the plural noun *sisters*. Therefore, the plural form of the verb (*come*) should be used.

5. **If WHO, WHICH, or THAT is used as a pronoun subject, the number depends on the noun for which the pronoun is substituting.**

Right This is the man *who talks* with you on the phone every day.
Wrong This is the man *who talk* with you on the phone every day.

Who is the subject of the verb *talks*. Since *who* substitutes for the singular noun *man*, the singular verb form (*talks*) is used.

Right These are the men *who talk* to you on the phone every day.
Wrong These are the men *who talks* to you on the phone every day.

Who is the subject of the verb *talks*. As *who* substitutes for the plural noun *men*, the plural verb form (without the *s*) is required.

REPLAY UNIT 37

Here is an opportunity to Replay what you studied in both Units 36 and 37. Circle the subject (which may be more than one word) of each verb. In the blank, write *C* if the verb agrees with its subject. Otherwise write the form of the verb that does agree with the subject. **Remember: Add s to verbs to make them singular unless the subject is** *you* or *I*.

1. Several groups was invited to participate in the meeting. ________
2. She don't want to go to the meeting with me. ________
3. There was several people waiting for the tickets. ________
4. Chan & Lopez, Inc., have an office in the new building. ________
5. The latest figures on yesterday's sale is available. ________
6. The report and the letter were on my desk. ________
7. A report on the accounts have been completed. ________
8. Neither he nor they are going. ________
9. Neither he nor his assistant wrap the parcels carefully. ________
10. Here are the new machines. ________
11. Miss Adams frequently arrive late. ________
12. The principal and his assistant drives to work together. ________
13. He usually do his work carefully. ________
14. The new typewriter don't work well. ________
15. My cousin, as well as many of my aunts and uncles, work here. ________

16. Neither the accountant nor the manager is able to prepare it.
17. Either Mr. Perez or I gives the test on Tuesday.
18. There go the latest models.
19. These are the typewriters which has new ribbons.
20. The class are finishing their application letters.
21. Roberta Simon, as well as all her aides, deserve a raise.
22. Another carton of envelopes has just arrived with the return address incorrectly printed.
23. Every man and woman in this country want to serve.
24. All the calendars was mailed to our customers yesterday.
25. One-third of the pies has been sold.

Please check answers on page 319.

READ UNIT 38

IF I WERE A MILLIONAIRE

We have all indulged in fantasies about what we would do if we were someone else or if conditions were enormously different from our present reality. The English language provides us with a special verb form for the unreal—that is, for expressing ideas that are contrary to reality. This special form—the word WERE—is used principally when the subject comes after *if, as though*, or *wish*. Use *were* regardless of what the subject is.

Was is ordinarily used with the following subjects: *I, he, she, it*, other singular pronouns except *you*, all singular nouns.

Ordinary	***Contrary to Reality***
I was at home yesterday.	If *I were* you, I would accept that position.
He was staring at the princess.	John wishes that *he were* a prince.
Everyone is (or *was*) *going to* the meeting.	If *everyone were* to go to the meeting, who would mind the store?

REPLAY UNIT 38

Make necessary corrections. If the sentence is already correct, write C in the blank.

______ 1. Even if the temperature were to reach 120°, I would not use the air conditioner.

______ 2. If he was in good health, we would offer him the job.

______ 3. Because he was ill so often last year, he couldn't complete his work.

______ 4. If he was a faster typist, he could complete that letter on time.

______ 5. I wish that I was your secretary instead of your wife.

______ 6. He was a millionaire who spent his money to help the poor.

______ 7. I wish that I was a millionaire.

______ 8. When I were your assistant, your office were in the other building.

______ 9. Cinderella wishes that she was a princess.

______ 10. If I was you, I would hesitate to make that promise.

Answers to this Replay are on page 319.
POP QUIZ—page 273.

Checkpoint

In the 1600s a mathematician and grammarian named John Willis was disturbed to discover that the words "shall" and "will" had just about the same meaning. He worked out a plan to give them different meanings, and defenseless students have been suffering ever since.

Those of you who studied grammar in the past may be wondering how we could go through this verb section without a will/shall–would/should unit. Well, the simple truth is that our ever-changing language patterns make this instruction unnecessary. While the distinctions between these pairs of words are still observed by some well-educated people, other good writers and speakers in business, politics, and the media are ignoring those grammar-book rules dating back to Mr. Willis.

Only the essential principles of verb usage are included in this section. Therefore, to be confident of your mastery of English for business and professions, make certain you have given adequate concentration to every item in Section Seven. If further details are desired, see the Grammar Supplement.

Repeat the correct forms aloud until they sound natural. For example, perhaps Unit 38 contains a principle you haven't been careful about following before. In that case, say "If I were you" five times every morning and every evening. Then, using your best theatrical technique, say "If I *was* you," making yourself sound ignorant and uneducated.

The most serious of the common verb errors are the following.

1. Using the present verb form when the past is needed:

   Yesterday Mrs. Jaffe talk to the business class about promptness. (should be *talked*)

2. Not adding *s* to a verb when it is needed or vice versa—adding *s* when it should not be there:

   Mrs. Oliver explain medical terminology every Friday. (should be *explains*)
   Mrs. Steffey and Mrs. Long teaches at El Camino College. (should be *teach*)

Special Assignment

1. Compose sentences with the present-time form of these verbs and any singular subjects except *you* or *I*.

Samples beat—*She beats* her husband instead of the eggs every morning.
throw—*Murphy throws* the ball too fast.

(a) go ______________________________

(b) do ______________________________

(c) drive ____________________________

2. Compose sentences using the *ing* form of these verbs. Use them as **verbs** rather than as some other part of speech.

Samples dance—*Don't*—My hobby is dancing. (*Dancing* is being used as a noun—your hobby can't dance around the floor.)
Don't—My husband attends dancing school. (*Dancing* is an adjective here.)
Do—My husband is dancing. (*Dancing* is a verb.)

(a) run ______________________________

(b) rise ______________________________

(c) forget ____________________________

3. Compose sentences using the appropriate form of these verbs to show past time **without a helping verb**.

Sample talk—*He talked* on the phone for an hour.

(a) win ______________________________

(b) sing ______________________________

(c) quit ______________________________

(d) burst _____________________________

4. Compose sentences using the appropriate form of these verbs **with a helping verb** to show past time.

Sample drink—The treasurer *had drunk* himself into a stupor.

(a) hide ______________________________

(b) pay ______________________________

(c) sink ______________________________

(d) freeze ______________________________

5. Compose sentences with the present-time form of these verbs and any plural subject.

Sample give—*They give* trading stamps with all purchases.

(a) cost ______________________________

(b) ship ______________________________

(c) equip ______________________________

This assignment should be completed by ____________. (date)

Proofreading for Careers

The names have been changed in this letter from a congressman to his constituent. In addition to name changes, you'll find errors, most of which were not in the original letter. I've added typing, capitalization, spelling, and grammar errors so that you can check your proofreading expertise. This time, however, there are also errors in the expression of numbers. Refer to "Numbers in Business" in the back of this book to be sure of writing numbers in the preferred style.

Glide your pen or pointing finger below each line as you proofread and correct the letter on the following page.

CONGRESS OF THE UNITED STATES
HOUSE OF REPRESENTATIVES
WASHINGTON, D.C. 20515

April 19, 198__

Mr. Eric A. Johnson
40711 Bendigo Drive
Monroe, LA 71201

Dear Mr. Johnson:

Thank you for you're correspondents concerning the nucular freeze initiative. As you no, the house has began debate on this issue.

All american's look forward to the day when Nucular Weapons are not a part of any country's military arsenel. You may wish to read the entire nucular freeze debate by securing copies of the March 16th and April 13th congressional records.

Additionaly, I have enclose 2 statements that expresses my views on this subject.

Thank you for sharing your concerns with me on this important issue. If you any further thoughts regarding this or any other issues, please feel free to contact me.

Sincerely,

Daniel J. Monson
Member of Congress

hs
enclosures

After verifying your proofreading and corrections, evaluate your proofreading skill: Excellent __ Good __ Fair __ Poor __

Practice Quiz

Refer to your dictionary whenever necessary. Circle the correct answer.

1. The Bureau of Mines (is/are) now making preparations to transfer its offices. (Unit 35)
2. The people of this country (believe/believes) it can be done. (Unit 36)

3. The manager has already (chose/chosen) three people for the job. (Unit 32)
4. The producers (employ/employs) most extras through the Central Casting Corp. (Unit 31)
5. At present 3,800 extras (is/are/be) on the roll of this bureau. (Unit 34)
6. If I (were/was) you, I would take advantage of this opportunity. (Unit 38)
7. Although the book (was/were) received some time ago, I have not read it. (Unit 36)
8. Mr. Hilton wrote me that I (did/done) the right thing. (Unit 32)
9. I wish that Ms. Watson (was/were) here to see these plans. (Unit 38)
10. You have (broke/broken) one of the rules. (Unit 32)
11. We believe that either Tom or James (is/are) to be fired. (Unit 36)
12. Each girl and boy (is/are/be) doing well. (Unit 36)
13. Accuracy in figures (mark/marks) the expert accountant. (Unit 37)
14. A list of names and addresses (come/comes) with the booklet. (Unit 37)
15. There (is/are/be) many subjects that you should master. (Unit 37)
16. Every one of the bookkeepers (attend/attends) night school. (Unit 36)
17. Either a letter or a pamphlet (was/were) sent. (Unit 36)
18. He (don't/doesn't) need to go. (Unit 32)
19. Have you (took/taken) notes of what he told you? (Unit 32)
20. Neither this month nor last month (has/have) been profitable. (Unit 36)
21. The systems analyst explained that the new HP computer (is/was) being used now. (Unit 34)
22. Both this year and last year (has/have) been profitable for us. (Unit 36)
23. Everyone who completes the Office Administration major (seem/seems) to pass the test. (Unit 36)
24. The results of this experiment in humanistic education (is/are) significant. (Unit 37)
25. In the rear of the word processing center (is/are) two new electronic typewriters. (Unit 37)

SECTION 7

FINE
GREAT
NEAT
FOXY
BEAUTIFUL
Smart
CUTE
WONDERFUL
INTELLIGENT
LOVELY

LOVE LINES

After Completing This Section, You Will:

- *Use singular and plural pointing adjectives correctly.*
- *Use a and an correctly.*
- *Use comparative and superlative adjectives and adverbs correctly.*
- *Avoid using double negatives.*
- *Distinguish between adjectives and adverbs and know when to use each.*

Men of all ages—before the current trend toward greater honesty in relationships—used to take pride in the "lines" they dispensed to the women from whom they wished to gain special favors. "Your graceful movements, pearl-like teeth, star-kissed eyes, petal-smooth skin . . ." They'd have been lost without adjectives and adverbs.

But how about us today? Can you imagine how dull our language and our lives would be without words like *generous, happily, stingy, cheerfully, prudish, confidently, weaker, meaner, strictest, domineering, shabby, purple, most comfortable* . . . ? Describing words—adjectives and adverbs—add precision, character, liveliness, and color to our language.

Communication is difficult at its best, for we are constantly misunderstanding one another. The skillful writer and speaker draws from a wealth of adjectives and adverbs, using those that will most effectively communicate the shade of meaning desired. In fact, our personalities are judged (subconsciously, of course) by the quality of our adjectives and adverbs.

In this section, you review the principles of Standard English for adverbs and adjectives.

VERY EASY

The adverb VERY is probably the most overused adverb in English. When an intensifying word is used too much, it loses its power. William Allen White, famous journalist and writer of the early 1900s, was editor of the *Emporia Gazette,* a well-written Kansas newspaper. To discourage the staff from overuse of VERY, he sent them this memorandum:

If you feel you must write "very," write "damn."

Since the copy desk had instructions to delete any of the reporters' profanity, the writing style of the newspaper was improved:

"It was a very fine victory" was written, "It was a damn fine victory," and was printed, "It was a fine victory."

READ UNIT 39

DEZE, DOZE, AND DEM

According to an AP news article, toiteen Noo Yawk cabbies got de foist verse lessons of der lives at a voice and diction class held in Nathan's Famous Restaurant on Times Square, a popular eating spot for New York's taxi drivers. Dis was a' attempt to help dem talk beda wid da ridin' public.

If you are not a New York cab driver, here are the only reminders *you* might need for correct use of pointing adjectives.

- Remember there are only four words to be used as pointing adjectives—this, that, these, those. (When used without a noun following them, they become pronouns.) The word "them" should never be used as a pointing adjective.

NO! I plan to give them boys a million dollars.
(*Them* is used incorrectly as a pointing adjective.)
OK I plan to give those boys a million dollars.
(*Those*—or *these*—is used correctly as a plural pointing adjective.)
OK I plan to give them a million dollars.
(*Them* is used correctly as a pronoun—a substitute for a noun naming some people.)

- Remember to use *this* and *that* as adjectives pointing to *singular* nouns only.

This kind of material is too sheer for keeping out the sun.
I don't like *that type* of person.
(*This* and *that* are singular pointing adjectives describing singular nouns *kind* and *type*.)

- Remember to use *these* and *those* as adjectives pointing to *plural* nouns only.

Those kinds of people are always complaining.
Will *these types* of homes be suitable in this neighborhood?
(Plural pointing adjectives: those, these. Plural nouns: kinds, types.)

- Remember that the "combinations" *this here* and *that there* and *them there* are nonstandard English.

NO! This here is my office.
OK This is my office.
NO! That there telephone is out of order.
OK That telephone is out of order.
NO! Them there paintings should be hung in the hallway.
OK Those paintings should be hung in the hallway.

REPLAY UNIT 39

Are you a "quick-change artist"? If so, correct the following sentences by the addition of a single letter or by the change of a single word.

1. The purchasing agent ordered those kind of calculators.
2. Them books should be returned to the library.
3. If you are not careful, these type of errors will occur frequently.
4. Those sort of books are extremely interesting.
5. I don't like those type of people.
6. He wants to buy this here book.
7. Please give that there calculator to the new clerk.
8. Tell them there people that they must do the work themselves.
9. That there is the way the cookie crumbles.
10. These kind of advertisements don't attract our customers.

Answers are on page 320.

READ UNIT 40

TWO LITTLE WORDS

Another kind of adjective is called an *article*. **The, a,** and **an** are the only three articles. *The* is known as a "definite" article because it makes the noun that comes after it definite or specific. *A* and *an* are "indefinite" articles. Notice the difference in meaning between "the book" and "a book."

People who are not natives of an English-speaking country have considerable difficulty learning when to use articles and when to omit them because no rules govern their usage. It's just a matter of knowing what "sounds right" to a native of an English-speaking country.

We are going to school today.

BUT

We are going to the office today.

Why don't we use *the* before "school" in the first sentence? Why *do* we use *the* before "office" in the second sentence? For foreigners learning English, the only solution is to read English a great deal and to ask teachers, friends, or relatives to correct you.

The other problem with articles is one that does affect native-born Americans—the two little words *a* and *an*. Some people use *a* almost exclusively and are not accustomed to using *an* when it is necessary for Standard English. Nonstandard use of *a* and *an* is very noticeable to business and professional people. Since there *are* useful rules to determine whether to use *a* or *an*, correcting this usage is much easier than the "article" problem that the foreigner has.

Not everyone needs to study the a/an rules. Some people automatically use them appropriately. Complete Replay 40 as a pretest; check your answers to determine whether you need to study all, some, or none of the rules that follow in Unit 41.

REPLAY UNIT 40

Write *a* or *an* in the blank before each of these words. Choose the answer that comes to you immediately. Don't try to figure out what the right answer should be.

1. ___adding machine	2. ___carrot	3. ___egg	4. ___apple
5. ___giant	6. ___honor	7. ___heater	8. ___English teacher
9. ___hand	10. ___island	11. ___manager	12. ___IBM typewriter
13. ___one-day visit	14. ___European	15. ___onion	16. ___FBI report
17. ___order	18. ___UN member	19 ___uncle	20. ___2 percent increase
21. ___11 percent drop	22. ___heir	23. ___X ray	24. ___unknown admirer

Check your answers on page 320.

READ UNIT 41

MANAGING "A" AND "AN"

If you had 100 percent accuracy on Replay 40, you do not need to study the rules that follow. Just go directly to Replay 41.

If you made errors in Replay 40, study these rules carefully and USE THEM AS A REFERENCE IN ANY WRITING YOU DO.

BASIC RULE FOR MANAGING "A" AND "AN"

If the word begins with a vowel sound (a, e, i, o, u), use *an*. Otherwise use *a*. The use of *a* or *an* depends on the *SOUND* OF THE WORD THAT FOLLOWS **a** or **an**, not necessarily the actual written letter with which the word begins.

an apple	a secretary
an Easter egg	a hat
an owner	a bookkeeper
an uncle	a 10 percent raise

- Words beginning with *u* sometimes begin with a consonant *sound*; for example, the *u* in *union* sounds like the *y* in *you*. Therefore, when a word begins with *u* that has the same SOUND as in *union* (the alphabet sound of *u*, sometimes called the long *u*), use **a** as the article.
- The *o* in *one* sounds just like the *w* in *winner*. Therefore, when a word begins with an *o* that has the sound of *w*, use **a** as the article.
- When *h* begins a word, it is sometimes silent—as in *honor*. The first sound of the word *honor* is *o*, a vowel sound; therefore, use **an** before a word that begins with a silent *h*.
- When these consonants are used by themselves or as abbreviations, notice that they begin with a vowel **sound**:

  F (ef) L (el) M (em) N (en) R(ar) S (es) X (ex) H (aitch)

  Since they begin with a vowel sound, use **an** before them. Before all other consonants, use **a** because the others begin with consonant sounds.

  C (se) D (de) K (ka) T (te) J (ja) Z (ze) etc.

For Example Insert **a T** or **an F** in the blanks.
He needs **an FBI** report for **a CIA** investigator.

Note Some abbreviations are pronounced as words instead of individual letters—such as NASA (pronounced nasa). He is **a** NASA employee. *If you're not certain of what NASA means, look it up NOW in your dictionary.*

> Use **a** before a consonant **sound**.
> Use **an** before a vowel **sound**.

REPLAY UNIT 41

Write *a* or *an* in each blank.

1. He was given __a__ one-day unpaid leave of absence and __an__ eight day paid vacation.
2. __a__ union member ate __an__ onion daily during __an__ 80-day tour of the world.
3. It is __an__ unusual combination for someone to be __a__ CPA and __an__ M.D. also.
4. They left for __a__ meeting about __an__ hour ago.

5. ____ thesaurus is ____ invaluable tool for those who work with words and ideas.
6. ____ X ray was needed to determine whether there was ____ injury.
7. ____ SEATO* representative and ____ NATO* representative were seated next to each other at ____ UN† meeting.
8. Flutter is ____ wavering in tape speed that can cause ____ unnatural sound.
9. Each time he received ____ B on ____ report card, he was given ____ $11 gift.
10. He walked down ____ hall to get ____ history book for ____ honest man.

Please check your answers on page 320.

READ UNIT 42

GOOD, GOODER, GOODEST?

If you've been around young children, you've heard such childish talk as "Mine is the goodest of all." Very soon, however, children learn, just from listening to others, to say "best." They don't need any rules. It just happens naturally. This unit reviews those aspects of adjective use that should be learned because they are *not* usually acquired without specific instruction.

When using "descriptive" adjectives (see Unit 3), you frequently want to show some comparison.

The secretary's typewriter is *newer* than the clerk's.
(Adjective *newer* compares one typewriter with another.)

Josephine is the *oldest* girl in the family.
(Adjective *oldest* compares the age of Josephine with the ages of the other girls in the family.)

DEGREES OF ADJECTIVES

Adjectives that may be used for comparisons come in three degrees:

1. Positive Degree—the entry word in the dictionary (good, bad, high, beautiful, etc.)
2. Comparative Degree—the form used for comparing just two (better, worse, higher, more beautiful, etc.)
3. Superlative Degree—the form used for comparing three or more (best, worst, highest, most beautiful, etc.)

*Pronounce SEATO and NATO as words: *seato* and *nato*.

†Pronounce UN as two separate letters.

GUIDELINES FOR MAKING COMPARISONS WITH REGULAR ADJECTIVES

An adjective describes a noun:

This NEW SHOW is at the Bijou Theater.
or
The SHOW at the Bijou Theater is NEW.

An adjective describes a pronoun:

IT is NEW.

A regular adjective has regular, or normal, changes in form for making comparisons. An irregular adjective does not follow any particular pattern for comparisons.

1. When comparing **two** nouns or pronouns, add **er** to one-syllable adjectives as well as to two-syllable adjectives ending with *Y*.

   (a) Jean is the *older* of the two sisters, but Marian is the *prettier*.*
   (Just two sisters are being compared.)

   (b) Your cookies taste *sweeter*, but mine are *bigger*.
   (Just two kinds of cookies are being compared.)

2. When comparing **three or more** nouns or pronouns, add **est** to an adjective with one syllable or with two syllables ending in *y*.

   (a) Melvin is the *oldest* child in that family, and he is also the *silliest*.
   (There are three or more children in that family.)

   (b) Yours is the *finest* diamond that I have ever seen.
   (I have seen three or more diamonds.)

3. When an adjective has two syllables *not* ending in *y* or has three or more syllables:

   (a) Use **more** or **less** before the adjective to compare **two** nouns or pronouns:
   The new calculator is *more expensive*.
   (Two machines are being compared.)
   The old equipment is *less valuable* than the new.
   (Two types of equipment are being compared.)

   (b) Use **most** or **least** before the adjective to compare **three** *or more* nouns or pronouns:
   Here is the *most recent* letter we have received.
   (We have received three or more letters.)
   The chiffonier is the *least valuable* of all the antiques.
   (Three or more antiques are being compared.)

*When the adjective ends in *y*, change *y* to *i* before adding *er* or *est*.

4. NEVER use *more, most, less,* or *least* before an adjective that ends with *er* or *est*:

HORRORS!!!

He is the most handsomest man on the screen today.
Mine is more better than hers.

GUIDELINES FOR MAKING COMPARISONS WITH IRREGULAR ADJECTIVES

When you think the comparison forms of an adjective differ from the ones just reviewed, refer to your dictionary. The irregular forms are next to the entry word, which would be the adjective in its ordinary form. For example, look up the adjective "good" right now. Next to it, you should find "better" and then "best" in that order. Other irregular adjectives are "far," "ill," "bad," and "many."

DICTIONARY DATA

When comparative and superlative degrees are shown in the dictionary (following the positive-degree entry word), they are presented in just that order. That is, the comparative form is always before the superlative, thus providing the sophisticated dictionary user with the information for choosing the correct form.

> Some one- and two-syllable adjectives are correct with either er/est *or* more/most. For example:
>
> CORRECT He is the handsomest man on the screen today.
> CORRECT He is the most handsome man on the screen today.
>
> When you're sure either one sounds right, use whichever you please. When in doubt, avoid error by following the rules given in this unit.

REPLAY UNIT 42

Part A

Look up these adjectives in your dictionary and fill in the blanks. Because they are irregular, you'll find the comparative and superlative forms next to the positive, which is the entry word.

1. far FARTHER or FURTHER (comparative) FARTHEST or FURTHEST (superlative)

2. bad	______________ comparative	______________ superlative
3. little	________ or ________ comparative	________ or ________ superlative
4. much	______________ comparative	______________ superlative
5. good	______________ comparative	______________ superlative

Part B

Correct any adjective errors in these sentences. Write *C* next to the one sentence that is already correct.

1. When you examine the two diagrams, you discover that the one on the right is biggest.
2. She lives farer away than her partner.
3. The most safest investments are in blue-chip companies.
4. Of the two reports, his is worst.
5. He is the younger of the two brothers.
6. He is the older of the three brothers.
7. When the work of the two accountants was compared, the treasurer found Mr. Grant's work to be accuratest.
8. This newly developed alloy is more heavier than any other metal.
9. This file contains recenter information than that one.
10. That trailer is the widest of the two that I saw.
11. Hers is the beautifulest office I have ever seen.
12. He feels worser today than yesterday.
13. She is the least popular of the two girls.
14. If you use our detergent and brand X, which one will give you the brightest wash?
15. He is the friendlyiest man I ever met.

Answers are on page 320.

Please take POP QUIZ on page 274.

READ UNIT 43

TO LY OR NOT TO LY, THAT IS THE QUESTION

That's not exactly the way Shakespeare wrote it, but it rhymes with the original anyway (if you pronounce lē and not lī). "Ly" is added to many adjectives to form adverbs. For example, adding *ly* to the adjective "occasional" results in the adverb "occasionally."

Most of the time we automatically know when to use an adverb and when to use an adjective. This unit is about only those parts of adjective versus adverb usage that most people need to study in order to use the correct form.

REMINDER:
(a) An adjective describes a noun or a pronoun.
(b) An adverb describes a verb, an adjective, or another adverb.

DISTINGUISHING BETWEEN ADJECTIVES AND ADVERBS

1. Most (but not all) words that end in *ly* are **adverbs**: *prettily, happily, quickly, busily.*

2. Some **adjectives** end in *ly* and should not be mistaken for adverbs: *curly, goodly, godly, friendly.* If the *ly* word in question can describe a noun (curly hair, friendly man), you'll know it's an adjective.

3. Sometimes the same word serves as either an adjective or an adverb:

   He is a *fast* worker.
   (*Fast* is an adjective describing the noun *worker.*)
   He works *fast.*
   (*Fast* is an adverb describing the verb *works.*)

4. Sometimes an adjective and adverb with similar meanings are completely different words:

   He is a *good* speaker.
   (*Good,* an adjective, describes the noun *speaker.*)
   He speaks *well.*
   (*Well,* an adverb, describes the verb *speaks.*)

5. Use the dictionary when in doubt about whether a word is an adjective or an adverb.

DETERMINING WHETHER TO USE AN ADJECTIVE OR AN ADVERB

1. Sometimes a descriptive word comes after an **existence verb**.

RECOGNIZING EXISTENCE VERBS

The most common existence verbs are forms of *to be*, such as *is, am, are, was, were, been*. Here is one way to determine whether a verb is an existence verb. See whether a form of the verb *to be* makes sense in place of the verb in question. For example:

She looks pretty **makes sense as** She is pretty **or** She was pretty.
The bosses seem happy **makes sense as** The bosses are (or were) happy.

Another way involves the verbs of the senses, which are sometimes existence verbs and sometimes action verbs. The most common verbs of the senses are *appear, become, seem, look, taste, sound, smell, feel*. Think of whether the verb is referring to action or to existence (state of being).

For example:

She *looks* for the eraser. (action verb)

This implies action—she is using her eyes to "look."

BUT

She *looks* sick. (existence verb)

This implies existence or state of being—she is not using her eyes to "look."

The mashed potatoes *taste* delicious. (existence verb)

Taste implies existence and not the action of "tasting." The potatoes are not "tasting."

BUT

First, I *taste* the mashed potatoes. (action verb)

Now *taste* implies action.

A describing word that follows an existence verb is an *adjective*; it describes the subject of the existence verb.

He looks *handsome*.
Handsome (which follows the existence verb) is an adjective describing the subject *he*.
The boss seems *happy*.
Happy (which follows the existence verb) is an adjective describing the subject *boss*.

2. A describing word following an action verb is an *adverb*; it describes the verb.

He arrived at the office on time and worked *quickly*.
Quickly is an adverb describing the verb *worked*.
Quick would be incorrect because an adjective should not describe a verb.

She knows English *well*.
Well is an adverb describing the action verb *knows*.
Good would be incorrect because an adjective should not describe a verb.

3. Since "real" and "sure" are adjectives, do not use them to **describe** other adjectives.

   She is *real* smart. (Wrong!)
   The adjective *real* cannot correctly describe the adjective *smart*; therefore switch to an adverb such as **really** or **extremely**.

   Remember that **adverbs** describe **adjectives**.

   That report is sure good. (Wrong!)
   The adjective *sure* cannot correctly describe *good*, which is also an adjective.
   Change to an adverb such as **very, certainly, exceptionally**, and so on.
   Only an adverb may describe an adjective.

4. Some frequently used adverbs are acceptable without the *ly* in very informal language, such as advertisements and road signs.

   Buy direct.
   or
   Go slow.

   These are not acceptable in most business writing.

IMPROVING ADVERB USAGE

1. Eliminate unnecessary adverbs:

AVOID	USE
repeat again	repeat
return back	return
cooperate together	cooperate

   The expressions on the left are redundant; that is, they are needlessly repetitive.

2. The WHERE words are often used as adverbs: anywhere, everywhere, somewhere, nowhere. Adding an *s* to any of those words results in a nonstandard usage.

SPELLING HELP IS HERE
The *i*ly ending always has one *l*—as in easily, busily, happily.
The a*ll*y ending always has two *l*'s—as in accidentally, occasionally, officially.

REPLAY UNIT 43

Part A

Concentrate on these directions; then follow one step at a time. All the sentences contain adjectives.

(a) Underline the adjectives that describe.
(b) Circle the word described by the adjective.
(c) If you circled a noun or a pronoun, write *C* (for correct) in the blank.
(d) If you circled a verb, cross out the adjective and write the appropriate adverb in the blank.

Sample She types careless. carelessly

1. He walked quick in order to meet her on time. ________
2. The fumes from the factory smell especially bad today. ________
3. Business people should write clear and correct. ________
4. Mrs. Adams looks beautiful. ________
5. However, her sister feels bad. ________
6. Be sure to see that you do the problems careful today. ________
7. The doctor wrote legible. ________
8. The pie tasted delicious. ________
9. The manager should think deep about that subject. ________
10. I hope that you will treat him fair. ________
11. This one works as good as the new one. ________
12. We sure wish you would attend the session. ________
13. I know English good. ________
14. She certainly speaks logical. ________
15. He appears calm. ________

Part B

In the blank, write the words or letter that should be omitted.

16. Please repeat that information again in your next speech. ________
17. We are returning your order back to you. ________
18. If we all cooperate together, the job will be done quickly. ________
19. I couldn't find the tool boxes anywheres. ________
20. No matter how hard he worked with the crew, he got nowheres. ________

Answers are on page 321.

READ UNIT 44

COMPARISONS

Shakespeare compared a woman to a summer's day in these lines:

Shall I compare thee to a summer's day?
Thou art more lovely and more temperate:
Rough winds do shake the darling buds of May.
And summer's lease hath all to short a date.
But thy eternal summer shall not fade.

In Unit 42, the *comparison* forms of adjectives are reviewed. The same three degrees of adjectives (positive, comparative, and superlative) are found in adverbs. Here are the principles that assure correct adverb comparisons.

For an adverb created by adding *ly* to an adjective, use *more* or *less* to form a comparative adverb and *most* or *least* to form a superlative adverb. As with adjectives, the comparative is for two and the superlative for three or more. In business and professional writing, do remember to use an adverb (not an adjective) in sentences like those shown below. In speech, the error of substituting an adjective for an adverb is not too noticeable and is preferable to fumbling while trying to think of the right word.

(a) The new machine works *more quietly* than the old one.
Notice that the adverb *quietly* is used to describe the verb WORKS.
Do *not* use the adjective *quieter* to describe a verb.

(b) Of all the businesses that have access to the parking lot, our company can obtain parking permits *most easily*.
Notice that the adverb *most easily* describes the verb OBTAIN.
Do *not* use the adjective *easiest* to describe a verb.

(c) He learns *more quickly* than the new assistant.
Notice that the adverb *more quickly* is used to describe the verb LEARNS.
Do *not* use the adjective *quicker* to describe a verb.

If you choose a synonym for "quick"— FAST—then you can't go wrong.
Fast is an adjective: He is a *fast* learner.
(*Fast* is an adjective describing the noun *learner*.)
Fast is also an adverb: He learns *fast*.
(*Fast* is an adverb describing verb *learns*.)

The comparative for both the adjective and the adverb is *faster*, and the superlative for both is *fastest*.

(CHALLENGE TO THE EXPERT: Think of a sentence using *fast* as a verb.)*

*To lose weight, you might fast for a day.

WELL, BETTER

He writes **better** than any other child in the class.
(The adverb *better* describes verb *writes.*)
If no comparison were made, the positive degree adverb would be used:

He writes **well**.

NOT

He writes **good**.

(*Good*, which is an adjective, cannot correctly describe a verb—*writes.*)

HOWEVER, the comparative of both *good* and *well* is *better*; the superlative is *best*. To those in the know (who read the last paragraph of Unit 42), this information is available in the dictionary.

Although *well* is an adverb, it may be used as an adjective for referring to **health**:

He feels *well.*
or
He feels *good.*

Either word correctly describes the pronoun *He*.

REPLAY UNIT 44

These sentences provide practice for Unit 44 as well as units 42 and 43. Alertness and concentration are required for correcting the adjective and adverb errors. If the sentence is correct, write C in the answer blank. Otherwise circle the adjective or adverb error and write the correct form in the blank. Use your dictionary whenever it might help.

1. Mr. Updyke is speaking more logical than anyone else in the room. ______
2. To learn quicker, study just one section at a time. ______
3. She appears more calmly than her sister. ______
4. His work was done the worstest of all. ______
5. I sure do like your office. ______
6. Ms. Teller speaks weller than anyone else on the Board. ______
7. Oscar runs faster than his brother. ______
8. Oscar also walks quicker than his brother. ______
9. I did real good in the grammar part. ______
10. The music on this boat sounds more beautifully than the music on the other boat. ______

11. He feels badder today than he did yesterday.
12. Mr. Allen, who is a good dictator, enunciates more distinct than does Ms. Stern.
13. The engine in the truck runs smoother than the one in the car.
14. She looks more efficient than the other candidates.
15. The prosecutor presented his case clearer than the other attorney.
16. Mr. Paige feels more badly about his failure to recognize Dwight than about anything else.
17. Of the five companies, Danton & Sons has the better standing.
18. Nureyev dances more graceful than the others in the troupe.
19. He protested the results the most bitterly of all the contestants.
20. He types the most satisfactory of all the applicants for the job in the word processing department.

Please match your answers with those on page 321.

READ UNIT 45

NOBODY DOESN'T LIKE SARA LEE

When two negative words are used to express one negative idea, the expression is known as a double negative.

> "I do*n't* want *no* spinach!" the child said.
> Then Mother rushed to the table with a big bowl of spinach and said, "If you don't want *no* spinach, you must want *some spinach.*"

The idea of a double negative equaling a positive was thought up by an 18th-century British bishop named Robert Lowth. He wrote a book called *Short Introduction to English Grammar* in which he intended to "lay down the rules" and to "judge every form and construction." He based the double negative rule on classical Latin despite the fact that the structure of English is different from that of Latin. He didn't even apply the Latin principle correctly, since double negatives *are* used in Latin for emphasis.

Actually, the use of two negative words to express a single negative idea does not in any way affect your listener's or reader's ability to understand your message. You know the child doesn't want any spinach. If your employer asks you what time it is and you reply, "I don't got no watch," she will certainly understand what you mean.

BUT

She will undoubtedly be reluctant to assign you to written or oral communications with her customers. In fact, using "double negatives" will often prevent applicants from being hired for white-collar jobs. The use of double negatives also affects the social standing of a person who wishes to socialize with business and professional people.

It isn't important to identify which negative words are being used as adverbs and which as adjectives. Just remember that two negative words should not be combined to express one negative idea. Examples of negative words that should not be combined are:

> *no, not, neither, never, none, nowhere, nobody, scarcely, rarely, barely, seldom, don't, doesn't, won't, can't, shouldn't, aren't, wouldn't, couldn't, haven't, hardly*

When two of the preceding words are combined, the result is one of the most obvious nonstandard usages in the English language. No matter how we may regard Bishop Lowth's rules, it's much easier to use single negatives than to risk losing job opportunities and even social opportunities.

THIS REALLY HAPPENED

A student showed me a punctuation quiz before turning it in. "I didn't put no comma there, Mrs. Smith," he said, pointing to one of the sentences. Trying to save him from the horrors of a life of double negatives, I said, "I didn't put *any* comma there, Joe." He replied, "Oh, good, you didn't put none there neither," thus increasing his original double negative to a triple.

If you suspect that you use double (or even triple) negatives, the best way to kick the habit is to say them aloud (to yourself) and make them "sound wrong."

REPLAY UNIT 45

Without changing the meaning or the basic sentence structure, correct these sentences so that they conform to Standard English. Write C next to the one sentence that is already correct.

1. I scarcely never do that anyway.
2. You don't need no fur coat in the summer in Southern California.
3. Don't put none over there as it might spill.
4. He won't eat no more pizza if he goes to China.
5. If you have a negative attitude, you won't succeed.
6. He don't know nothing about chemistry.
7. Jonathan won't go nowheres with me on Saturday.

8. A good student wouldn't never need to cheat.
9. If she eats enough fruits and vegetables, she shouldn't get sick no more.
10. The manager doesn't need no secretary.
11. He ain't never in when you want to see him.
12. He hardly never saw the stock.
13. You couldn't hardly expect Mr. Blank to join such an organization.
14. There's hardly no difference between Wahl's and Kimble's estimates.
15. Nobody doesn't like Sara Lee Cheesecake.

See page 321 for answers.
Turn to page 275 and take the Pop Quiz.

DID YOU KNOW THAT . . .
The PRINCIPAL in colonial America was the "principal" teacher in a school that had two or more teachers. The principal's job was to keep the records and see that the schoolhouse was kept clean.

DEAN was originally the title given to a superior over ten monks. The word comes from the Latin "decanus," which was derived from the Latin word for "ten" (decem).

Checkpoint

Adjectives include the four pointing words: this, that, these, those. Just remember that the first two are singular and the latter two are plural.

Another type of adjective is the *article*. There are three articles: **the, a, an**. Use *a* before a word beginning with a consonant sound and *an* before a word beginning with a vowel sound.

Limiting adjectives that "limit" nouns in the sense of quantity, as explained in Section One, are usually used correctly and are therefore excluded from study in this section.

Describing adjectives come in three degrees: Use the positive degree when nothing is being compared; that is, for a description of just one item. The comparative degree (usually ending in *er*) is used when two nouns or pronouns are compared by the adjective. For comparing three or more, the superlative should be used and is usually constructed by adding *est* to the adjective. When *er* or *est* can't be added, use mo**re** or mo**st** (or less or least) before the adjective.

Adverbs frequently end in *ly*. Use adverbs, not adjectives, to describe verbs, adjectives, and other adverbs. However, remember that a describing word that follows an existence verb is an adjective; it describes the *subject* of the existence verb.

Adverbs are used for making comparisons in the same way as adjectives. When the adverb ends in *ly*, use *more* (or less) before it for the comparative form and *most* (or least) for the superlative.

Negative words are either adjectives or adverbs. It isn't important whether you identify them as such. Just be sure to use just one negative word to express one negative idea.

> **WHICH SENTENCE REFERS TO A SOCK?**
> It's darned good.
> It's darned well.

> **WHICH DOG WOULD DO POORLY IN POLICE WORK OR HUNTING?**
> The dog smells bad.
> The dog smells badly.

If you wish additional details about adjectives and adverbs, see the Grammar Supplement in the back of the book.

Special Assignment

Compose a report of 120–150 words (12–15 typewritten lines) comparing two people you know who differ from each other in some way. Choose a specific difference to write about—such as appearance, personality, character, intelligence, or ability. "Introduce" both people and your subject in the opening sentence. Use simple and concrete language.

SAMPLE OPENING SENTENCES

These are poor:

(a) I am going to write about . . .

(b) This is a comparison of two people I know who differ from each other.

These are good:

(a) Johnson and Stern both make important but different contributions to the success of the basketball team.

(b) Although Dr. X and Dr. Y are both good instructors, their teaching methods differ greatly.

(c) It's hard to believe that June and Joan are twins because their personalities are so different.

(d) Ellen is a perfect example of how a supervisor should dress for the office, while Janice is just the opposite.

Body Continue with sentences that "support" your opening statement. Describe the people with interesting, accurate, and precise words.

Ending Conclude with a summarizing type of sentence. Avoid introducing a new subject with your closing sentence.

Submit your report on ____________. (date)

Proofreading for Careers

Use your proofreading skill to detect and correct spelling, grammar, typographical, number, and abbreviation errors. See the "Reference Section" in the back of this book for information about numbers, capitalization, and abbreviations. Always proofread the entire letter, not just the body. Use your pen or pencil as a pointer and mark errors as you go along.

IDAHO POTATO GROWERS EXCHANGE
POCATELLO DRIVE
BOISE, IDAHO 83700

March 6th, '85

Ms. Maryann Lamb, Editer
Health Magazine
1300 5th St.
Wenatchee, Wash. 98801

Dear Ms. Lamb:

In response to you Feb. 27 letter, we are pleased to share the following facts about potatos with you:

Potatos is composed of 78% water, about eighteen per cent carbohydrate, and about two percent protein. Their a good source of iron and vitamines.

It is a erroneous idea that poatoes is fatening. Nutritionists recognize that all foods eaten in excess are fattening. Actualy potatoes are less fattening than most items in the daily american diet. Potatoes alone are more lower in calories per pound than bread and many other foods. When fryed or served with butter and sour cream; however, the total caloric intake is high.

Potatoes has a high satiety value and imparts a full feeling that checks overeating. Sodium content is so slight that the American Heart Association reccomend potatoes for low-salt diets.

The enclose pamplet was wrote by Nancy Borden who lives in Southwick, a area known for their potato industry. Also enclosed are 2 recipes; we think the Potatoes Granada recipe would be the best for your publication.

Please let us know if we can be further help to you.

Sincerly,

Mike Rothstein

lr
enclosures

After checking with the corrected version of this letter, evaluate your skill.
Excellent __ Good __ Fair __ Poor __ Other? __

Practice Quiz

In the blank, write the letter that identifies the correct answer. Use the dictionary when in doubt.

Sample (a) A (b) An eager history professor assigned 200 pages. b

1. (a) A (b) An union official would probably (Unit 41)
2. refuse (a) a (b) an hourly wage. (Unit 41)
3. Of the two designs we have considered, the first seems (a) best (b) better (c) more better. (Unit 42)
4. Of all the colors we have considered, brown is (a) more appropriate (b) most appropriate (c) appropriatest (d) appropriater. (Unit 42)
5. She doesn't want to lose (a) a (b) an $82 commission. (Unit 41)
6. If he doesn't understand the work, he will get (a) a (b) an F on the test. (Unit 41)
7. (a) A (b) An European businessperson would probably (Unit 41)
8. consider this award to be (a) a (b) an honor. (Unit 41)
9. Since both brands are equally good, please order the (a) less (b) least expensive. (Unit 42)
10. Which is (a) easier (b) easyer (c) easiest (d) easyest for you to type, a manuscript or a tabulation? (Unit 42)
11. Mr. Vork is a (a) faster (b) more faster typist than Mr. French. (Unit 42)
12. I would prefer to have Harold rather than Richard assist me, as Harold is (a) patientist (b) most patient (c) more patient (d) patienter. (Unit 42)
13. Her work is even (a) badder (b) worse (c) worser (d) worster than her sister's. (Unit 42)

Cross out adverb or adjective errors, and write the correct word or words in the blank. If the sentence is correctly written, write C in the blank.

Sample Mrs. Ellis walked ~~slow~~ to the store. slowly

14. The class was asked to sit quiet anywheres. (Unit 43)
15. The flowers smell good. (Unit 43)
16. I hope you won't do no more work on that job. (Unit 45)
17. Ms. Lupino spoke brief and to the point when she addressed the club. (Unit 43)
18. The new copying machine operates smoother than the old one. (Unit 44)

19. Of all the stores in the city, this one is managed the poorest. (Unit 44)

20. He furnished the office different from what I would have expected. (Unit 43)

21. The chemicals don't smell as bad today as they did yesterday. (Unit 43)

22. The business machine show isn't open no more, but it will reopen again in May. (Units 43 & 45)

23. It seems that Dr. Berman wrote this prescription clearer than usual. (Unit 44)

24. The accountant and the auditor were real unhappy about the figures. (Unit 43)

25. This report was typed the most accurate of all the reports. (Unit 44)

SECTION 8

THE TAMING OF THE APOSTROPHE

After Completing This Section, You Will:

- *Distinguish between possessive nouns and nonpossessive plurals.*
- *Write possessive nouns correctly.*
- *Write contractions correctly.*
- *Use the apostrophe correctly in special plural forms.*
- *Use the apostrophe symbol correctly for expressing feet, minutes, years, and certain quotations.*

The shrew (a scolding, nagging woman) in Shakespeare's *Taming of the Shrew* didn't know "her place" and didn't know how to behave as a dutiful wife until her husband "tamed" her—a concept quite unacceptable today.

Our concern, however, is with the taming of the apostrophe. When apostrophes are untamed, they appear in the wrong places and are missing from the right places. The result is writing that is unclear or distracting.

It is important to learn exactly where to use apostrophes in order to avoid the common error of using them where they don't belong. Take advantage of the opportunity to master the use of this little mark. Since the apostrophe is all you concentrate on in this section, you'll always be confident in the future of your ability to place the mark correctly.

WHICH BUTLER WOULD GET FIRED?
The butler stood at the doorway and called the guests names.
The butler stood at the doorway and called the guests' names.

READ UNIT 46

THE UBIQUITOUS "S"

It's here again. But then that's what "ubiquitous" means—seeming to be everywhere at the same time. You're probably not surprised to know that the letter *s* is used to begin more words than any other letter in the alphabet and is the second most frequently used consonant (*t* is the first).

In Section Four you worked with the ubiquitous **s** while it played the original numbers game—plurals. The **s** is also used to make nouns possessive. For that reason, be careful to avoid confusing plurals with possessives. In addition to an **s**, possessive nouns require an apostrophe,

BUT

nouns that are **simply plural** and **not possessive** are written without an apostrophe.

The conscientious student, aware of the need for apostrophes, sometimes sprinkles them like raindrops wherever an **s** happens to end a word. Studying this section ensures you will *not* be an apostrophe sprinkler and you *will* use apostrophes where needed.

DISTINGUISHING BETWEEN PLURALS AND POSSESSIVES

POSSESSIVE NOUNS

Possession in grammar terminology shows a relationship between one noun and another. The relationship is made clear to the reader by use of an **s** and an **apostrophe**. Various kinds of relationships are included in the possessive category.

Personal—Mary's brother; the manager's friend.
Ownership—the men's hats; a girl's pen.
Place of origin—Nebraska's population; New York's weather.
Authorship—the president's speech; the accountants' report.
Type or kind—children's clothes; ladies' fashions.
Time—two months' delay; a week's vacation.

Whether to place the apostrophe before or after the *s* is explained in the next unit.

When a possessive relationship exists between two nouns, use either the apostrophe method as shown above or use extra words. A possessive noun thus replaces a prepositional phrase to show a possessive relationship. Compare the prepositional phrases below with the shortcut method in the box above.

the brother of Mary; the friend of the manager
the hats belonging to the men; a pen belonging to the girl
the population of Nebraska; the weather in New York
the speech given by the president; the report prepared by the accountants
a delay of two months; a vacation of one week

The sentences below show how a group of words can be shortened by the use of a possessive noun. Studying these sentences enables you to know when to make a noun possessive. Notice that it is always the **first** of the two nouns that is made possessive.

Prepositional Phrases (without apostrophe)	***Possessive Nouns*** (with apostrophe)
The *records prepared by the accountants* were taken to the *office of the secretary.*	*The accountants' records* were taken to the *secretary's office.*
The population of Nebraska is smaller than the *population of New York.*	*Nebraska's population* is smaller than *New York's.*

Clothes for children are on the fifth floor, and *clothes for infants* are on the fourth floor.	*Children's clothes* are on the fifth floor, and *infants' clothes* are on the fourth floor.
Delay of two months would be disastrous to this project.	*Two months' delay* would be disastrous to this project.

In the second sentence on the right, "population" is understood but not stated after the possessive "New York's."

To find out whether an apostrophe is needed, try to reverse the order of the nouns and put "of" between them:

Marys brother—brother of Mary

If the reversal and the insertion of "of" works, then the original expression is possessive and the first noun needs an apostrophe.

Mary's brother

PLURAL NOUNS

Most plural nouns end in **s**. Just because a noun ends in *s* doesn't mean that an apostrophe is required. **Use the apostrophe only for a possessive relationship**. These sentences have plurals that end in *s*; no possessive idea is indicated. Apostrophes would be INCORRECT.

The *brothers* are *partners* in a plumbing supply business.
The *girls* each own several *books*.
The *ladies* will wear straw *hats*.
Your *records* are inaccurate.
Several new *accountants* are needed in our office.
Three of the *secretaries* go to lunch at one o'clock.
All the *factories* that they own are in small *towns*.
In two *months* we hope to start selling the new merchandise.

When one noun that ends in *s* is followed by another noun, the first noun is usually possessive. Notice that the nouns ending in *s* in the sentences above are not directly followed by other nouns. Hence they are not possessive and do not require apostrophes.

REPLAY UNIT 46

Some of the nouns in this Replay are possessive. The apostrophes, however, are left out. First underline the noun that is "possessed" or "owned." Then circle the noun that is the "possessor" or "owner" (the word that needs an apostrophe). Please do not insert apostrophes; just circles. Write C for correct beside the sentences that don't require an apostrophe.

Sample The (salesmans) <u>books</u> were left in (Mr. Foxs) <u>office</u>.

C 1. The Williamses have sent four sopranos to try out for the chorus.

2. The new editors stories pleased his readers greatly.

3. His brothers-in-law manage the office.
4. His brother-in-laws manager has been transferred to the Chicago offices.
5. The Joneses own several pieces of property in the swamp lands of Brazil.
6. All the attorneys offices are in the new buildings.
7. South Dakotas resources are listed in the back pages of the Almanac.
8. Mens and womens clothes are on sale in all the stores today.
9. Have you shipped Mrs. Smiths orders yet?
10. One of the film industrys talented directors, Steven Spielberg, will deliver several lectures at UCLA.
11. The crews strength was spent in a useless maneuver.
12. Even at their peak, the gold mines in California were less profitable than the orange groves.
13. Californias vineyards are the source of more than 75 percent of the nations wine and almost all of its raisins.
14. In three months we hope to see McCarthys greatest invention.
15. Claremont Mens College is one of about fifty privately supported institutions of higher education in Southern California.
16. Oral communication in business includes making introductions, giving direction, greeting visitors, and making announcements.
17. The secretarys instruction book was used for several weeks before the errors were noticed.
18. Several weeks work was completed in less than three days.
19. The teachers and parents inspected the results of the fire that destroyed nearly half the Kavanaugh School Districts classrooms.
20. We are all looking forward to Ellens visit during Easter vacation and hope she will enjoy the sights of Southern California.

Answers should be checked now on page 322.

READ UNIT 47

BEFORE OR AFTER?

Several hundred years ago some writers showed possession by using the pronoun "his" after the first noun. Instead of writing *the clerk's desk*, they wrote *the clerk his desk*. If you say this old-fashioned possessive form fast, you'll find you can hardly hear the first two letters of "his." Therefore, people of that day, when spelling was much more individualized than it is today, began spelling the expression *the clerk s desk*. Since the apostrophe had previously been used to show that one or more letters had been left out of a word, writers began to use this mark to show that the first two letters had been left off "his." That's how we ended up with the modern form, *the clerk's desk*.

If more than one clerk shared the desk, the original expression would have been *the clerks their desk*. This, in turn, was shortened to *the clerks' desk*—the apostrophe coming after the *s* to show that the word "their" was left out after the plural word "clerks."

You can still use this simple way of determining whether the apostrophe goes before or after the *s* in a possessive noun. Once you've determined that the noun is possessive as explained in Unit 46, then decide whether it is singular or plural. If it is singular, put the apostrophe where the word "his" would have been in the 1700s. If it is plural, insert the apostrophe where the word "their" would have been.

Although the "his" and "their" trick works every time, here are some simple rules to give you added confidence that you are forming possessives correctly.

RULE 1—SINGULAR NOUNS

Add *'s* to spell the possessive of a SINGULAR noun.

(a) The *secretary's* desk is on wheels. (one secretary)
(b) We do not believe the *witness's* testimony. (one witness)
(c) Please send the *accountant's* report to the controller. (one accountant)
(d) The *city's* needs for extra funds have not been met. (one city)
(e) Mr. *Gaines's* office is around the corner. (one Mr. Gaines)

BUT

If adding apostrophe+*S* to a *singular* noun makes the word hard to say, just add an apostrophe. Use this alternate method when the noun (usually a proper noun) has two or more syllables and ends with an *s* or a *z* sound.

(a) *Mr. Stettinius'* wife was not active in Washington social life.
(b) *John Alberts'* office is on the main floor.
(c) *Moses'* journey is described in the Bible.
(c) *Steinmetz'* inventions made him famous.

RULE 2—PLURAL NOUNS ENDING IN *S*

Add only an apostrophe to spell the possessive of a PLURAL noun that ends in *s*.

(a) The *secretaries'* desks are on wheels. (more than one secretary)
(b) The *witnesses'* statements are false. (more than one witness)
(c) Send the *accountants'* report to the controller. (more than one accountant)
(d) Throughout the nation, the *cities'* needs have not been met. (more than one city)
(e) The *Gaineses'* home is in Sioux City. (more than one person named Gaines)

RULE 3—PLURAL NOUNS NOT ENDING IN *S*

> Add *'s* to spell the possessive of a *plural* noun that does not end in *s*. Notice that each word *in italics* is plural but does not end in *s*.

(a) *Men's* suits are very colorful this year.
(b) The *children's* assets are held by the trustee.
(c) The *alumni's* contributions were small.

Apply this same rule to plural compound nouns. As you studied in Section 4, *sons-in-law* is the plural form. The word does not end in *s* but in *w*.

(d) Your *sons-in-law's* business has gone into bankruptcy.

Note: Rules 1, 2, and 3 require that before placing the apostrophe, you decide whether the noun is **singular** or **plural**. Then make certain the noun is correctly and completely spelled before making it possessive.

Special Case When the name of an organization includes a possessive form, just see how the organization writes it. As in spelling a person's name (Stephen/Steven), there is no right or wrong.

These two organizations do not use apostrophes in their names:

Silverwoods (department store)   Columbia Teachers College

These two do use apostrophes:

Ladies' Home Journal   I. Magnin's (department store)

No Exceptions **Men's, women's, children's, man's, woman's**, and **child's** are always written with **'s** when they show possession.

REPLAY UNIT 47

Part A

Write the singular possessive and the plural possessive.

		SINGULAR POSSESSIVE	PLURAL POSSESSIVE
Sample	*lawyer*	lawyer's	lawyers'
	1. freshman	FreSHMAN'S	Freshmen's
	2. week	week's	weeks'
	3. witness	witness's	witnesses'
	4. James	James's	Jameses'
	5. country	country's	countries'

6. secretary
7. clerk
8. mouse
9. wife
10. father-in-law
11. congresswoman
12. manager
13. fox
14. hour
15. Wolf
16. wolf
17. principal
18. boss
19. lady
20. child

Part B

Insert an apostrophe and an *s* wherever needed to show that a noun is possessive. Show clearly whether the apostrophe is before or after the *s*. Three sentences do not require any apostrophes.

If the grammar and meaning indicate that a noun should be **plural** but **not possessive**, add **s** only, and make any necessary spelling change. Remember that subjects and verbs should agree in number. Read each sentence carefully before deciding on the correction.

1. The boy sweater was torn during the football game.
2. Several new member were initiated into the club.
3. Each of the community has a civil defense program.
4. Two year interest is due on the note.
5. A person health is his most valuable possession.
6. I think men fashion change as quickly as women.
7. This new company asset are limited.
8. The two editor in chief remark were helpful.
9. Be prepared to come at a minute notice.
10. Mr. Perkins signature was needed two day ago.
11. The store was having a sale on lady coat.
12. The Goldstein of Palm Beach will join us in two day.
13. San Diego population has increased during the past five year.
14. Mr. Jones desk is to your right.

15. The chairman report included several detail on the proposed worker cafeteria.
16. We studied Keats poetry in our literature class.
17. Montreal and Quebec are in Mr. Knox territory.
18. Fritz brother is chairman of the board.
19. I don't believe Hendrix story.
20. The butler stood at the doorway and called all the guest name.

Check answers on page 322.
Turn to page 276 for Pop Quiz on Units 46–47.

READ UNIT 48

MORE APOSTROPHES

Here are additional uses for the apostrophe.

CONTRACTIONS

In the dictionary section you observed how changing the accent mark for a word can change not only the pronunciation but also the meaning. The word CONTRACT, for example, can be pronounced with the accent on the first syllable to mean an agreement. When the accent is on the second syllable, it means to reduce in size by drawing together. Hence the word CONTRACTION refers to a word that has been shortened by removing one or more letters. It also means joining two or more words together and leaving out a letter or two. The apostrophe is placed at the exact point where the letters are omitted. (A period should not be used at the end of a contraction unless it happens to end the sentence.)

Here are some words frequently contracted in business writing. Just look at them carefully and remember that the apostrophe belongs where the missing letter or letters would have been.

is not=isn't	are not=aren't	was not=wasn't
were not=weren't	have not=haven't	would not=wouldn't
should not=shouldn't	could not=couldn't	will not=won't
cannot=can't	has not=hasn't	do not=don't
does not=doesn't	of the clock=o'clock	I have=I've
you have=you've	we have=we've	I shall or will=I'll
I will=I'll	you will=you'll	we shall or will=we'll
I am=I'm	he is=he's	you are=you're
we are=we're	they are=they're	that is=that's
what is=what's	it is=it's	she is=she's

These contractions are acceptable in ordinary business writing and speech. Whenever the contraction sounds natural to you, use it. However, in extremely formal writing, avoid contractions.

Contractions other than the types listed above—such as *nat'l* for *national, sec'y* for *secretary*—are used only in informal notes or in tables where it is necessary to save space.

PLURALS

Use an apostrophe for the plural of letters and words.

(a) Please be sure to dot your *i*'s. (plural of letter)
(b) The applicant's college transcript showed four F's in various math classes. (plural of letter)
(c) Please omit all "therefore's." (plural of word)

An apostrophe is unnecessary (but not wrong) when forming the plural of numbers or capital-letter abbreviations. For **small-letter** abbreviations, do use the apostrophe.

(a) The temperature in Akron is in the 70s.
(b) All C.O.D.s should be sent to my office.
(c) There are two YWCAs in Toledo.
(d) All c.o.d's should be sent to my office. (plural of lowercase abbreviation)

When an abbreviation is **possessive**, use the apostrophe—the same as with any other noun.

(a) The AMA's position is clear. (singular possessive)
(b) The R.N.s' uniforms are yellow at this hospital. (plural possessive)

MISCELLANEOUS

As a symbol, the apostrophe has several different meanings. Here are the most common.

Feet—4' Minutes—4'	Use only in technical work, tables, charts, and so on; otherwise, spell out the word.
Century—'84	Except for class graduation years and well-known historical years, do not use the apostrophe to replace the century in business letters and formal reports. Instead, use the numbers: 1984.
Quotation within a quotation	The candidate said, "It was Abraham Lincoln who referred to 'government of the people, by the people, and for the people'; and that is also my credo." Use an apostrophe to type the single quotation marks required for a quotation within a quotation.

REPLAY UNIT 48

Insert apostrophes where needed. Write C next to the sentences that do not require the addition of an apostrophe.

1. I couldnt meet you at five oclock.
2. Please be sure to dot your is.
3. All C.O.D.s should be referred to the receiving manager.
4. In the 1920s America was prosperous and optimistic about the future.
5. Its predicted that the 1990s will be an era of full employment.
6. Dont forget that the R.N.s uniforms are yellow.
7. Several M.D.s have their offices on the third floor.
8. There are too many ands, ifs, and buts in your composition.
9. The class of 65 held its reunion at the finest hotel in the city.
10. There were many suicides following the stock market crash of 29.
11. There are three CPAs in this building.
12. The Gay 90s was a period when workers suffered while the wealthy had lavish parties.
13. If you cross your ls, theyll look like sloppy ts.
14. Wont you please marry me?
15. Only five As were recorded.
16. In the early 1900s, women wore long dresses.
17. Whats the difference between Stock No. 234 and No. 235?
18. Youve only two forms left.
19. Arent you going to hire that accountant?
20. Couldnt you make your 9s look less like 7s?

Answers are on page 322.

Checkpoint

Tame apostrophes, just like pets, obey the rules and are in the right place at the right time. After completing the Practice Quiz, Proofreading, and the Special Assignment according to the "rules" of Units 46–48, you'll be sure your apostrophes are ready for company—any company you work for.

Test your mastery of the apostrophe principles just studied by placing a check next to each one you feel sure of. If your conscience will not permit the check mark at first, reread the explanations in the unit.

______ Apostrophes are used in singular and plural possessive nouns. (Unit 46)

______ If a plural noun isn't possessive, don't use an apostrophe. (Unit 46)

______ BUT to form the plural of a small-letter abbreviation, a single letter, or a word "used as a word," use an apostrophe. (Unit 48)

______ A possessive noun ends with *s* and comes before another noun—Charley's aunt. (Unit 46)

______ To test for possessives, reverse the order of the two nouns and insert "of" between them—aunt of Charley. (Unit 46)

______ A singular noun is made possessive by adding *'s*. (Unit 47)

______ If adding *'s* to a singular noun results in a possessive that is awkward to pronounce, the *s* may be omitted. When in doubt, add the *s*; you won't go wrong. (Unit 47)

______ A plural noun that ends in *s* is made possessive by adding an apostrophe only. (Unit 47)

______ A plural noun that does not end in *s* is made possessive by adding *'s*. (Unit 47)

______ An apostrophe is inserted within a contraction at the exact place where letters are left out. (Unit 48)

______ The apostrophe is the symbol used for feet, minutes, centuries, and quotes inside of quotes. (Unit 48)

If you'd like further details about the use of the apostrophe, see Section 8 of the Grammar Supplement in the back of the book.

Special Assignment

A. In Replay Unit 46, you were asked to circle the words that require apostrophes. List those words below and insert the apostrophes in the proper places. When no change was required in a sentence, just write "None" in the blank.

Sample salesman's, Fox's

1. __________
2. __________
3. __________
4. __________
5. __________
6. __________
7. __________
8. __________
9. __________
10. __________
11. __________
12. __________
13. __________
14. __________
15. __________
16. __________
17. __________
18. __________
19. __________
20. __________

B. Write sentences in which you shorten these phrases by using a possessive noun.

Sample son of Mr. Ames ______ Mr. Ames's son is the auditor.

1. books of Patricia ______ PrATICIA'S BOOKS
2. wife of Mr. Jones ______ Mr. JONES'S WIFE
3. vacation of one week ______ WeeK'S VACATION
4. home of the Adamses ______ ADDMBES' home
5. books belonging to the boys ______ The boys' BOOKS
6. toys of the children ______ the children's TOYS
7. problems of the members ______ members' Problems
8. commissions of the salespeople ______ salespeople's commissions
9. work of two years ______ TWO years' WORK
10. office of my father-in-law ______ FATHER'S-IN-LAW OFFICE
11. words of Moses ______ MOSE'S WORDS
12. streets of Houston ______ Houston's streets
13. report of the auditor ______ AuditoR'S report
14. notice of ten minutes ______ Ten minutes' notice
15. expense accounts of the supervisors ______ SUPERVISORS' expense accounts

C. Correct the following sentence:

Student's who put apostrophe's into plain plural's may receive shock's when they get grade's on the examination's in a few day's.

This assignment should be completed by ______. (date)

Proofreading for Careers

This short essay requires the addition of eight apostrophes. Also be on the lookout for two misspellings of nouns you studied in Section 4, two capitalization errors, and one comma splice.

In the worlds most famous museum, the louvre in Paris, hangs a painting by Americas celebrated artist, James McNeill Whistler. This paintings formal title is "An Arrangement in Gray and Black," but it is better known by the simple name "Whistlers Mother."

Many studies have been made in an effort to explain the basis of this portraits almost universal appeal. But what criterions can the art Critic use in judging a painting?

The critic is not like a scientist, he or she cannot set up a controlled experiment wherein a number of stimulus are shot into a subject and data collected on the subjects reaction. No, the art critic must rely on inner emotions and sensitivity when analyzing a painting.

Analyses of a painting is very personal. If you looked at "Whistlers Mother," what would you see? Would you, like most of us, be left wondering about the source of this portraits greatness?

After verifying your corrections, choose an adjective to evaluate your skill today: ________________________. Do you need to review apostrophes? _____ Section 4 (singular and plural)? _____ Comma Splice? _____ Capitalization (Section 4 and page 284)? _____. All OK? _____

Practice Quiz

After reading each sentence, indicate the correct forms of the words listed below the sentence.

Sample The chairmans report included details on the proposed workers cafeteria.

Answer	
b	1. (a) chairmans (b) chairman's (c) chairmans'.
a	2. (a) details (b) detail's (c) details'.
c	3. (a) workers (b) worker's (c) workers'.

The camp directors view was that drastic changes had to be made in Johns outlook. (This camp has one director.)

______ 1. (a) directors (b) director's (c) directors'. (Unit 47)
______ 2. (a) changes (b) change's (c) changes'. (Unit 46)
______ 3. (a) Johns (b) John's (c) Johns'. (Unit 47)

It was agreed that new desks should be installed in the salesmens office.

______ 4. (a) desks (b) desk's (c) desks'. (Unit 46)
______ 5. (a) salesmens (b) salesmen's (c) salesmens'. (Unit 46)

Yesterdays techniques cannot succeed in todays market.

______ 6. (a) Yesterdays (b) Yesterday's (c) Yesterdays'. (Units 46, 47)
______ 7. (a) techniques (b) technique's (c) techniques'. (Unit 46)
______ 8. (a) todays (b) today's (c) todays'. (Unit 46)

You have one weeks time to accept or reject this companys offer.

______ 9. (a) weeks (b) week's (c) weeks'. (Units 46, 47)
______ 10. (a) companys (b) company's (c) companie's (d) companys' (e) companies'. (Unit 47)

The managers, at last Wednesdays meeting, agreed to rebuild the executives recreation room.

______ 11. (a) managers (b) manager's (c) managers'. (Unit 46)
______ 12. (a) Wednesdays (b) Wednesday's (c) Wednesdays'. (Unit 47)
______ 13. (a) executives (b) executive's (c) executives'. (Unit 47)

I havent read the minutes of the meeting yet.

______ 14. (a) havent (b) have'nt (c)haven't (d) hav'ent. (Unit 48)
______ 15. (a) minutes (b) minute's (c) minutes'. (Unit 46)

The PTAs paper supply is in the top drawer.

_______ 16. (a) PTAs (b)PTA's (c) PTAs'. (Unit 48)

The presidents of five PTAs were invited to the reunion of the class of 75.

_______ 17. (a) presidents (b) president's (c) presidents'. (Unit 46)
_______ 18. (a) PTAs (b) PTA's (c) PTAs'. (Unit 48)
_______ 19. (a) 75 (b) '75. (Unit 48)

My sister-in-laws manager left college after two years work.

_______ 20. (a) sister's-in-law (b) sister-in-law's (c) sister-in-laws' (d) sisters'-in-law. (Unit 47)
_______ 21. (a) years (b) year's (c) years'. (Unit 47)

Miller and Jones is one of the citys finest firms.

_______ 22. (a) Jone's (b) Jones (c) Jones'. (Unit 47)
_______ 23. (a) citys (b) cities (c) city's (d) cities' (e) citie's. (Unit 47)
_______ 24. (a) firm's (b)firms (c) firms'. (Unit 46)

Houstons streets are safe at night.

_______ 25. (a) Houstons (b) Houston's (c) Houstons'. (Units, 46, 47)

DOORMATS IN SUBURBIA

A stroll through any middle-class suburb reveals incorrect apostrophes on doormats, mailboxes, and even in elaborate wrought iron signs. Notice the unneeded apostrophes in the names that identify the owners of the homes:

The Smith's The Jone's The Anderson's

To form the plural of proper nouns, merely add *s* or *es*:

The Smiths The Joneses The Andersons The Foxes

SECTION 9

THE PAUSE THAT REFRESHES

After Completing This Section, You Will:

- *Use commas exactly where they belong.*

When you are talking with someone, brief pauses between certain words help the listener follow the ideas you are expressing. If you don't pause at exactly the right places, the listener lets you know he's not sure of what you mean. Then you repeat or even change the order of the words to make sure your listener understands.

Writing, however, should be immediately understood by the reader since he cannot easily ask for a repeat or rephrasing. When you are writing, punctuation marks replace the pauses.

Good writers have developed a system for placement of these pausing symbols—or punctuation marks. Although punctuation through hearing (placing some mark wherever it seems there should be a pause) is helpful, it is too inexact a method by itself for a person writing a business letter, a report, instructions, and so on. When you master the rules in this section and the next, you no longer will rely on a kind of guesswork. Instead, your ability to punctuate by hearing will be increased by your understanding of the logical rules that follow. This section explains the system established for the most commonly used punctuation mark—the comma.

A WARNING As you complete these replay units, be sure to insert a comma only when the rule applies. If you punctuate without at least mentally referring to the rules each time, you will continue to make whatever comma errors you made in the past. In short, you will be wasting your time doing this work if you don't get the practice in applying the rules.

READ UNIT 49

WINE, WOMEN, AND SONG

Johann Voss, an 18th-century poet who may have been a bit of a reprobate, wrote:

> Who does not love wine, women, and song
> Remains a fool his whole life long.

"Wine, women, and song" is an example of a series. A series consists of three or more words, word groups, or complete thoughts.

(a) He filed all the letters, reports, invoices, and purchase orders. (words)
(b) We believe in government of the people, for the people, and by the people. (word groups—or phrases)
(c) Lani does the typing, Robert does the filing, and Dawn keeps the books. (complete thoughts)

Notice the commas between the items in the series as well as before the conjunction (*and* or *or*) that precedes the last item.*

If, however, the conjunction is used before *each* item in the series, commas are not used:

Engineers and physicists and chemists are working on the project.

REPLAY UNIT 49

Apply the comma rule for series by inserting commas in these sentences; if you are tempted to use a comma for any other reason, DON'T! If no comma is needed, write C in the blank.

______ 1. The company supplies gas electric light and power to a number of industries towns and cities.

C 2. Our agent and our accountant and our broker are all needed for consultation before this decision can be made.

______ 3. We looked under the desk in the trash and everywhere else we could think of but still couldn't find the report.

______ 4. First we studied the report then we investigated its accuracy and finally we decided to buy the stock.

______ 5. Chevrolets Fords Plymouths and VWs have always been lower priced than Chryslers Cadillacs and Lincolns.

______ 6. A meeting will be held on June 12 August 6 or September 19.

C 7. He doesn't care whether the uniforms are supplied in green or yellow or purple.

C 8. The job of a top manager is to set broad objectives formulate strategies to meet them and decide among alternate possibilities.

______ 9. Then he or she must determine which task should be performed first second and so forth.

*Business English experts recommend the use of that final comma, although you'll find that some literary writers and journalists omit it.

_____ 10. International Business Machines General Motors American Telephone and Telegraph and several other major business organizations have recently become slightly more autocratic in their internal relationships.

See page 323 for answers.

READ UNIT 50

FOR ADJECTIVES ONLY

The "series" rule of Unit 49 applies to the use of commas when there are *three* or more items in the series. However, with **adjectives**, use a comma between just *two* or more of them when "and" is omitted but understood.

They died during the long, severe winter.
(The adjectives *long* and *severe* both describe *winter*. "And" would make sense between them; therefore, use a comma instead of "and.")

They died during the long and severe winter.
("And" is between *long* and *severe*; therefore, no comma is needed.)

She is an intelligent, loyal employee.
(A comma replaces "and" between *intelligent* and *loyal*.)

She is an intelligent and loyal employee.
("And" is between *intelligent* and *loyal*; therefore, a comma isn't needed.)

Sometimes one word really describes two other words; in such cases, "and" does not make sense between the adjectives, and no comma should be used.

Many elderly people died during the long, severe winter.
(*Many* describes *elderly people*; you would not insert "and" between *many* and *elderly*; therefore, no comma is needed.)

Betty Van Meter drives a yellow sports car.
(*Yellow* describes *sports car*; "and" would not be logical between *yellow* and *sports*; therefore, a comma should not be used.)

REPLAY UNIT 50

Apply the comma rules for "adjectives only" and for series by inserting commas where needed. When you are tempted to use a comma for any other reason, DON'T! If the sentence needs no comma, write C in the blank.

_______ 1. She is wearing a new red dress and looks clean and neat.

_______ 2. Her pleasant friendly personality is an asset to this office.

_______ 3. He asked why you came what you wanted and what you expected to do.

_______ 4. Our office is located in a small elegant building in the new part of town.

_______ 5. Eric Smith is a quiet and efficient worker and is also an excellent photographer.

_______ 6. Susan Jaffe will travel with you to Paris Brussels and Munich.

_______ 7. We need a courteous helpful friendly salesman like Evan Simon in this business.

_______ 8. Harley will be healthy wealthy and wise.

_______ 9. Payment will be made by certified personal check.

_______ 10. The Middle East sheikdom of Bahrain has a settled well-educated population.

Answers are on page 323.

READ UNIT 51

INDEPENDENTS DAY

No, the title isn't spelled wrong, but then neither is it July 4. This unit reviews Section Two instruction about complete thoughts—independents.

- Use a comma between two complete and independent thoughts joined by *and, but, or, nor,* or *for.* Place the comma *before* the conjunction.

  The clerk filed copies of all the outgoing letters, but she forgot to file copies of the invoices. (Two complete thoughts are joined by "but.")

  The clerk filed copies of all the outgoing letters but forgot to file copies of the invoices. (Since a complete thought, or independent, does **not** follow "but," no comma is used.)

I am proud that you are one of us, and I want you to know that I appreciate your work. (Two complete thoughts are joined by "and.")

I am proud that you are one of us and want you to know that I appreciate your work. (Since a complete thought, or independent, does **not** follow "and," no comma is used.)

- When the two complete thoughts are very short (no more than about ten words), the comma is usually omitted if **and** or **or** is the joining word.

  Did she type or did she file?
  Francine wants the job and she will do it well.
  Ms. J. Mack is a member, **but** she didn't attend.

REPLAY UNIT 51

Apply the comma principles from Units 49–51 to these sentences. If you are tempted to insert a comma for another reason, RESIST! Write C next to the two sentences that do not require any commas.

____ 1. I have thought the problem over carefully and I shall give you the answer in an hour.

____ 2. Alan Berman spoke with Amy today but he couldn't persuade her to do the research.

____ 3. Progressive Drug Company has never done any banking with us but we have heard it has a good reputation.

____ 4. Mrs. Jaffe will teach the shorthand class in the mornings but is unable to teach any afternoon classes this year or next.

____ 5. A well-educated person is needed to manage the new marketing division and I think Mr. Chandra might be just the man for the job.

____ 6. Either you may pay 20 percent of the price as a down payment or you may pay the entire amount now.

____ 7. Doris lives in Rochester but she visits Spokane often.

____ 8. Lucille Podmore is a creative conscientious teacher and has taught at several elementary schools.

____ 9. Are the summers warmer in Miami Fort Dade or Jacksonville?

____ 10. An advertising agency acts as an intermediary between a company that wants to advertise and the various media that sell space and time.

Check your answers on page 323.

READ UNIT 52

CURTAIN RAISERS

A curtain raiser is a short play presented before the principal dramatic presentation. The "curtain raiser" for a sentence is a group of words presented before the main part of the sentence. It is known as an **introductory expression** and is always a **dependent** group of words; that is, it cannot stand alone as a complete thought any more than the theatrical curtain raiser can be presented without the main show. The audience would want their money back.

Many, but not all, introductory expressions are separated from the main idea of the sentence by a comma.

1. Place a comma after an introductory expression that contains any form of a verb.* The verb form in these introductory expressions is underlined.

   (a) Seeing his chance for success, Mr. White accepted the assignment immediately.
   (b) If you attend the meeting, be sure to take notes for Ms. Todd.

2. Place a comma after an introductory expression that contains five or more words. (It doesn't matter whether or not it contains a verb or verb form.)

   (a) Because of the unusual circumstances, we shall allow you to enter the contest.
   (b) Under sponsorship of the University of Pennsylvania, a conference on hypnosis and psychosomatic medicine will be held on July 2.

3. Place a comma after the following introductory expressions: *Yes, No, Well, Oh.*

   Yes, we'll be glad to ship ten dozen toasters to you.

4. Place a comma after an introductory expression used to "talk" to someone. This is called "direct address."

   Joelene, will you help me?

5. When an introductory expression has fewer than five words and has no verb form, use your judgment to determine whether a comma is needed.

   (a) For a while, longer skirts will be worn. (Notice how the reader might be confused if the comma were omitted.)
   (b) In a month we'll know the results. ("time" introductory)
   In Alaska the winter nights are long. ("place" introductory)
   (Commas are not usually required after "time" and "place" introductories that are short.)

*Do not use a comma if the verb form (verbal) is the subject; for example, *To know you* is a joy.

REPLAY UNIT 52

Insert commas where needed. Apply the rules included so far in Section Nine. If tempted to use a comma for some other reason, be strong—DON'T! Write C next to the four sentences that need no commas.

______ 1. If you place your order now you'll receive the goods in ample time for the sale.

______ 2. In our attractive modern office on Wall Street you will find a courteous staff at your service.

______ 3. As we all know real merit is hard to conceal.

______ 4. Since you have not seen the report you and your staff should not be criticizing it.

______ 5. You'll receive the goods in ample time for the sale if you place your order now.

______ 6. Being an alert salesperson he noticed the prospect's gesture of annoyance.

______ 7. No he hasn't called on us either this month or last.

______ 8. Mrs. Lee we are unable to repair your washer dryer toaster and coffeepot all in one day.

______ 9. Please write a note to the conference leader if you cannot attend.

______ 10. If you need a new accountant call our agency.

______ 11. In a month Steve will receive his degree and will be ready to work as an engineer.

______ 12. Working quickly he carefully organized the information.

______ 13. Although the furniture arrived on time telephone service was not immediately available.

______ 14. Francine and Evan think Emerson would be a good location for the new factory but the client feels it would be too far from the home office.

______ 15. In Chicago many people sunbathe on the cement beach near Lake Michigan.

Answers are on page 323.
POP QUIZ—page 277.

READ UNIT 53

PARENTHETICALS

A parenthetical expression adds an extra idea to the sentence but doesn't change the main idea of the sentence. (You can picture the expression in parentheses.) When the parenthetical expression "interrupts" the main idea, commas are used before and after. Naturally, if the parenthetical expression is at the beginning or the end of the sentence, only one comma is needed. When the parenthetical expression is at the beginning, it might also be called an introductory expression. It does not matter what you call it, as long as you use the comma.

If removing the expression from the sentence would make the meaning unclear or would change the meaning of the sentence, it is not a parenthetical expression; do not use commas. Reading the sentence aloud may help to determine whether you want your reader to pause.

- Use commas in these sentences because the expressions are parenthetical (not essential). Each parenthetical expression is in italics (slanted letters). If you cover the parenthetical words, the rest of the sentence keeps its original meaning.

1. Our former neighbor is, *of course,* an expert in this field.
2. Herb Greenland, *who is an expert in this field,* will lead the discussion.
3. Herb Greenland, *an expert in this field,* will lead the discussion.
4. He is an expert in this field, *however,* and will lead the discussion.
5. *In conclusion,* several features of this proposal are unsatisfactory to us and need to be rewritten.
6. We shall definitely meet you at 5 p.m., *although we cannot remain for long.*

- When a person's name is used in **direct address**—that is, to talk to him or her directly—the name is parenthetical; use commas. Notice how you pause before the name when saying the sentence aloud.

7. We're sorry that we cannot pay you for delivering the lecture, *Mr. Greenland.*
8. Because of your expertise, *Mr. Greenland,* we'd appreciate your leading the discussion.
9. *Herb,* we'd appreciate your leading the discussion because of your expertise in this field.
10. Now, *ladies and gentlemen,* I'll show you a simple experiment in physics.

- When the parenthetical name is not a direct address, use commas if the name consists of more than one word.

11. My former neighbor, *Pat Greenland,* is a talented artist.
12. My former neighbor, *Mrs. Greenland,* is a talented artist.

BUT

- If just the first name is used, commas are not generally required. Notice that you don't pause before the name when saying the sentence aloud.

  13. My former neighbor *Pat* is a talented artist.
  14. My sister-in-law *Bea* will lead the discussion.
  *15. Your old friend *Herb* will meet you at the reunion.

- If a name is not parenthetical, do not use commas.

  16. I believe that *Professor Joanne Murcar* lives in Spokane.
  17. *Professor Chan-Nui* is from Wenatchee.

- When the title of a publication is parenthetical—or nonessential—use commas.

  18. Bergerud and Gonzalez' latest book, *Word Processing: Concepts and Careers*, is being studied in colleges throughout the country.

BUT

- When the title just flows into the rest of the sentence—or is essential—do not use commas.

  19. Have you read the book entitled *Word Processing: Concepts and Careers?*

- In the next group of sentences, commas are *not* used around the questionable expressions because they are essential (not parenthetical). Cover the italicized words; then notice how the meaning of the rest of the sentence is distorted.

  20. A person *who is an expert in this field* should lead the discussion. (Compare with Sentence No. 2.)
  21. Someone *like you* should lead the discussion.
  22. The document *that is dated June 12* should supersede all others.
  23. Will you please bring your résumé with you *when you come to my office?*
  24. A national organization *with a research and service bureau* stands behind our product.
  25. Any pedestrian *who doesn't obey the caution signals* should be fined.
  26. A city *that can qualify as a depressed area* is eligible to receive state and federal aid.
  27. Mr. Sampson is an instructor *who has time and patience to offer students needing extra help.*

*The constructions in No.'s 11–15 are often referred to as "appositives," or "nouns in apposition."

REPLAY UNIT 53

Insert commas according to the principle explained in Unit 53 only. Use your judgment to decide on the main idea intended by the writer of these sentences. Write C beside the eight sentences that are already correct.

______ 1. In the Chicago office however this plan saved the company a great deal of money.

______ 2. The operating costs as he probably told you are too high.

______ 3. Two of our employees Mr. Jones and Ms. Brown are attorneys.

______ 4. My office Room No. 105 is in this building.

______ 5. The Credit Department which is on the third floor will be open all day.

______ 6. Martin Simon the new accountant is responsible for the errors.

______ 7. The woman wearing a red dress is my sister.

______ 8. The boys who stole the bicycles were arrested.

______ 9. Your work therefore is better than anyone else's.

______ 10. We cannot accept this proposal Ms. Ross.

______ 11. Shari and Peter will lead the discussion today even though they are not experts.

______ 12. The new equipment which I haven't even learned to operate yet is out of order.

______ 13. Ted Banta will lead the discussion tomorrow although we probably should have asked an expert.

______ 14. Roberta's friend Marie works at the Bank of America.

______ 15. The merchandise will we believe be shipped by the end of next week.

______ 16. George isn't that Stella Glitter signing autographs in the lobby?

______ 17. Coaches who fail to inspire their players should be replaced.

______ 18. My sister-in-law Bette lives in Riverdale.

______ 19. The textbook we use in the office administration course is called *Machine Transcription and Dictation.*

______ 20. Openness fosters growth and establishes an important feeling of worth which motivates everyone to do the best possible job.

______ 21. Gerald Siegel is the attorney who will try the case.

______ 22. This instruction booklet which gives you complete information can be purchased for $3.95.

_____ 23. Barry Hicks enjoys effective and correct use of the English language, although he is not a purist.

_____ 24. Our team, consisting of five engineers, will design the new pipe.

_C_ 25. The book *In Praise of English* was written by Joseph T. Shipley.

Please check your answers on page 324.

READ UNIT 54

THREE EASY COMMAS

The three principles of comma usage in this unit are easy. Just about all you have to do is read through the unit and complete the Replay; then you'll be sure of these commas.

STATES

Almost everyone remembers the comma between city and state. It's the one *after* the state that some people forget. Use commas before and after the state when it follows the name of the city.

I lived in Springfield, Massachusetts, when I was young.
The art festival in Laguna Beach, California, attracts many tourists.

DATES

(a) Another principle that almost everyone remembers is the comma between a date and a year. But when a date is used within a sentence, a comma is needed *after* the year as well.

The meeting was held on April 5, 1980, in Canton.

(b) When a specific day of the month isn't used, however, commas are optional. Use both or none.

The meeting was held in April 1976 at the main office.
The next meeting will be in May, 1988, at the branch office.

(c) Enclose a date that explains a preceding day. Be sure to use commas both **before** and **after** the date.

On Wednesday, January 3, we shall begin the training sessions.

ABBREVIATIONS WITH NAMES

(a) Use commas before and after college degrees that follow a name.

Melanie Sweet, M.D., and Fran Smith, Ed.D, have been elected to the Board of Directors of City of Hope Hospital.

(b) Commas usually set off *Jr., Sr., II, III, Inc.,* or *Ltd.*

Charles Davis, Jr., and his son Charles Davis, III, have worked for Metromedia, Inc., for the past five years.

unless the person or company chooses to omit the commas.

Charles Davis Jr. and Charles Davis III worked for Avco Inc. before they joined Metromedia, Inc.

REPLAY UNIT 54

Insert commas where needed. Write C next to the three sentences that don't need a comma. If you are tempted to use a comma but can't think of the reason for it, DON'T!

1. We received your order on March 9 1984 and acknowledged it the same day.
2. Mara Simmons started to work for us in April 1984 and has been with us ever since.
3. Steven Smith Ph.D. is in charge of the Industrial Arts Department in the Saratoga School District.
4. Billy Crystal started his career as a comedian in a Long Beach New York nightclub.
5. The deposition was taken on Friday January 4 1984 with all witnesses present.
6. We hope to attend the Mardi Gras in New Orleans next year.
7. Little Rock Arkansas was the scene of tragedy and strife in the 1960s.
8. Goldfinger's Variety Stores Inc. closed the White Plains New York store in 1973.
9. Windsor Woolens, Ltd. carries mostly imports from Edinburgh Scotland and from London.
10. The director will begin interviewing on February 5 for the job of sales manager.

Answers are on page 324.

READ UNIT 55

NOTES ON QUOTES

NOTE 1 Use a comma to separate a direct quotation from the rest of the sentence.

Some people place an apostrophe directly over an *s* because they don't know whether it goes before or after. Another kind of fence sitter places quotation marks directly over a comma. The correct placement is very easy to learn by reading Note 2.

NOTE 2 Always place a comma or period *before* the quotation mark, never after—no exceptions!

(a) **Quote at end of sentence:** When the meeting began, he calmly said, "I am going home."
(b) **Quote at beginning of sentence:** "Money is the root of all evil," he replied.
(c) **Quote at middle of sentence:** He whispered, "I love you," and then fainted.
(d) **Quote that is interrupted:** "Money," he explained, "is the root of all evil."

NOTE 3 The word "is" in the sentence above shows that a lowercase letter is used to begin a quotation that is just a portion of a sentence.

NOTE 4 No comma is used when a quotation ends with a question mark or an exclamation mark:

"Will you arrive early?" asked Miss Alvarez.
"I don't believe it!" exclaimed Mr. Ripley.

NOTE 5 No comma is used when the quotation is not a complete sentence and is woven in with the remainder of the sentence:

We all sang "Happy Birthday" when Barbara entered the room.
Jenny Smith answered the boss's question with a simple "no."

REPLAY UNIT 55

The best way to remember the five notes of Unit 55 is to compose sentences in which you apply them.

1. Compose sentences that include the following:
   (a) a quote at the end of the sentence When classes began, he said, "I am not feeling well".

(b) a quote at the beginning of the sentence ____________________

(c) a quote at the middle of the sentence ____________________

(d) a quote that is interrupted ____________________

2. Now write a sentence beginning with a quotation that is a question or an exclamation.

____

____

3. Should commas be added to the sentences that follow? Yes ______

No ______ Maybe ______

(a) My teacher has told me over and over again to "type rapidly."

(b) Scarlett O'Hara wanted to do everything "tomorrow" in *Gone With the Wind.*

Please check your answers on page 324.

READ UNIT 56

KISS ME ONCE AGAIN, BUT NOT NOW

"Kiss me once, kiss me twice, and kiss me once again" is a line from a romantic ballad of the 1940s. Many activities do improve with repetition. Sometimes repetition of a word results in an improved sentence because of added emphasis.

- Use a comma between two identical words that appear together.

    Many, many years ago Eric became interested in tropical fish.
    He who cries, cries alone.

- Use a comma before a contrasting or opposing expression. These expressions are generally introduced by *not, never, seldom,* or *but.*

    Be sure to send for his brother, not his sister
    He has often considered leaving his job, never his wife.
    Kiss me once again, but not now.

> PUNCTUATION BLOOPER FROM NEWSPAPER ADVERTISEMENT
> House for rent. View takes in five counties, two bedrooms.

REPLAY UNIT 56

Insert commas where necessary. If you cannot apply one of the rules presented in Units 49–56, don't use a comma.

1. He arrived at 10 p.m. not a.m.
2. He was a quiet efficient worker.
3. I want to urge you however not to worry.
4. On June 16 1985 he was 65 years old.
5. No you are wrong.
6. "I know" he said "that you will be there on time."
7. I am very very happy that you received the promotion.
8. He works out at the gym every morning never in the afternoon.
9. If you decide to run run fast.
10. Although the quality is excellent the price is too high.

Answers are on page 324.

READ UNIT 57

A COMMA MEDLEY

A medley usually refers to a musical arrangement that includes several different melodies. In other words, it's a mixture. This unit is a medley of comma rules.

- A comma replaces omitted verbs and other important words that are easily understood from the way the rest of the sentence is worded.

  A used machine costs $1,245; a new one, $2,300.

  (The comma replaces the verb COSTS. Remember that a semicolon is used between independent clauses that are not joined by *and, but, or, nor*, or *for*—Unit 12.)

- A comma separates thousands, millions, billions when the number refers to a quantity. Starting from the right, count by threes and then insert the comma.

  $2,000 167,823 widgets 1,321,000 shmitchiks

  However, commas are not used in numbers that "identify," such as addresses, serial numbers, page numbers, and so on.

  page 1247 19721 Victory Boulevard No. 23890

- Commas separate the parts of an address that appear in sentence form.

  They had lived at 1114 Fteley Avenue, Bronx, New York, for many years.
  Please return the gold charm bracelets to Robert Stern, Inc., 7 Tappan Terrace, Ardsley, New York 10502, before next week.

  (A comma should never be used before the ZIP.)

REPLAY UNIT 57

All needed commas are correctly inserted in these sentences. On the blank line below each sentence, briefly state the rule (in my words or yours).

1. If you need more information, please let us know.
   ____
2. Please send the equipment to our West Coast Distributor: Stan 'n Ev Fashions, 328 North Palm Drive, Palm Springs, California 90036.
   ____
3. Our home office is at 6413 Third Avenue, but we have offices all over the world.
   ____
4. Please ship 167,823 shmitchiks and weblows to us at once.
   ____
5. Carol, Eric, and Harley ran up the stairs, along the hall, and into the office.
   ____
6. In September we hired a new secretary; the following month, a typist; and in November, a bookkeeper.
   ____
7. Geographic data for the three years ending October 31, 1982, is on page 1465.
   ____
8. After eating, my brother washed the dishes.
   ____
9. The novelist Ernest Hemingway once lived here.
   ____
10. No, you have the wrong number.
    ____

Answers to this Replay are on page 324.
A POP QUIZ on Units 53–57 is on page 278.

Checkpoint

Commas are written signals that, when correctly used, are almost as effective as a speaker's pauses and voice inflections. Misplaced commas or omitted commas can make a sentence difficult or impossible to understand.

A college student sent a telegram to her parents asking whether she could buy a fur coat that was on sale for $2,500. The parents intending to refuse the request, sent a telegram with the message: NO COST IS TOO GREAT. The missing comma after "NO" turned out to be an expensive mistake.

Here is another example of how the use of commas can completely change the meaning of some sentences.:

Bradley thinks Ronnie is very generous.
Bradley, thinks Ronnie, is very generous.

In addition to providing the writer with a method of making meaning clear, the correct use or omission of commas is an important segment of standard written English—because it happens in every sentence you write.

Place a check in the blank next to each comma rule you clearly understand. Review the ones you're not sure of.

______ 1. Use commas between the items in series as well as before the conjunction that precedes the last item. (Unit 49)

______ 2. Use a comma between two adjectives when *and* is omitted but understood. (Unit 50)

______ 3. Use a comma before *and, but, nor, or, for* when one of those conjunctions is used for joining two complete and independent thoughts. (Unit 51)

______ 4. Use a comma after an introductory expression: (Unit 52)

______ (a) with a verb form in it

______ (b) with five words or more

______ (c) when necessary for clearness

______ (d) consisting of **yes, no, well**, or **oh**

______ (e) when addressing a person by name

______ 5. Use commas around nonessential, or parenthetical, words. (Unit 53)

______ 6. Use commas before and *after* a state name that directly follows the name of a city. (Unit 54)

______ 7. Use commas before and after dates that explain preceding days. (Unit 54)

______ 8. Use commas before and after abbreviations following names—unless the person or company named doesn't want a comma. (Unit 54)

_______ 9. Use commas to separate a direct quotation from the rest of the sentence unless the quotation ends with an exclamation mark or a question mark. (Unit 55)

_______ 10. Use a comma before a word that is a repeat of the previous word when a pause or emphasis is desirable. (Unit 56)

_______ 11. Use a comma before a contrasting or opposing expression. These often begin with a word such as *but, seldom, never,* or *not.* (Unit 56)

_______ 12. Use a comma to replace omitted words. (Unit 57)

_______ 13. Use commas to separate thousands, millions, billions when the number refers to a quantity. (Unit 57)

_______ 14. Use commas between parts of addresses that appear within sentences. (Unit 57)

Grammar explanations for some of the comma rules appear in Section 9 of the Grammar Supplement in the back of the book.

Special Assignment

Insert commas in the sentences below. Always think of the rule before inserting the comma. Four sentences don't require commas. Complete this assignment by _______________. (date)

1. Mr. Lupino thought the problem over carefully and will give you the answer tomorrow. (Unit 51)
2. I talked with Ann Rosenberg today and you should receive a call from her by Thursday. (Unit 51)
3. We received your letter yesterday Mr. Chin and will ship your order by the end of the week. (Unit 53)
4. Shortly before eight thirty men appeared. (Unit 52)
5. As I have already been over the instructions I shall not repeat them. (Unit 52)
6. Feeling the way you do you should stay home for the rest of the day. (Unit 52)
7. "A man's best friend is his ape" said Tarzan. (Unit 55)
8. The teacher said "The most important asset is the will to study." (Unit 55)
9. "The law" said the speaker "is broken every minute of the day." (Unit 55)
10. If you will take the matter up with our coach Mr. James Heffron you will be given full information. (Unit 52)
11. There is a good opportunity of course to market our stationery envelopes and business forms to your most exacting trade. (Units 49, 53)

12. Mr. Kuwahara if you decide to take out any life insurance in the future the premium will be more than you are now paying. (Unit 52)
13. We believe that the owner The Robinson Realty Company will accept $750. (Unit 53)
14. Each alteration omission or addition costs us a great deal of money. (Unit 49)
15. We shall temporarily suspend service on your telephone 345-6789 from December 1 1988 until further notice. (Units 53, 54)
16. Henry F. Albert Jr.* of Cincinnati Ohio is the author of the new book. (Unit 54)
17. The publisher's price on this new book *America's Supreme Opportunity* is $18.95. (Unit 53)
18. The word "accommodate" has two *c*'s and two *m*'s. (Unit 53)
19. White and fancy colored leather shoes are selling fast but how are your customers keeping their delicate footwear clean? (Units 50, 51)
20. The assessment for March 2 1985 amounted to $26; for May 2 $32; and for June 4 $35. (Unit 54)
21. Automobiles that do not have brakes are hard to stop and should not be driven in Chaosville after midnight June 31 1989. (Units 53, 54)
22. Yes I expect to meet Inez Valdez who is the president of Premier Electronics Inc.* at two o'clock on March 6. (Units 52, 53, 54)
23. We are very very sorry that you suffered from such a long hot summer in Panama. (Units 50, 56)
24. We expect to receive this order by the first of October not December. (Unit 56)
25. Adam Clay the manager believes that no one is exempt from this ruling. (Unit 53)
26. Because you visited the office and talked with Mr. Chang we will expect a brief report not a book. (Units 52, 56)
27. Upon hearing the evidence he felt she was guilty. (Unit 52)
28. Phyllis Ross formerly of 2245 Prospect Avenue Bronx New York 10923 is now offering interior design service from the Miami Beach address shown below. (Units 53, 57)
29. In two or three more weeks the secretaries will be on vacation.
30. In some ways you cannot understand her viewpoint.
31. Mr. Kelly if you have already received the goods no reply to this letter is necessary.
32. This practice I believe is bringing about results that could not be obtained in any other way.
33. Samantha falls in love with every Tom Dick and Harry.
34. In Seattle the large amount of rain results in lush foliage.
35. During 1974 my family and I began to meditate twice daily.

*Henry (No. 16) and Premier (No. 22) use the commas.

Proofreading for Careers

As you proofread this letter to an Eagle Scout, correct commas, apostrophes, spelling, sentence construction, grammar, capitalization, and typing.

> Dear Fellow Eagel:
>
> Thank your for extending your membership in the National Eagle Scout association.
>
> There have never been a time in the history of our Country when *Scouting was needed more. Scouting needs it's Eagle Scouts, regardless of age to maintain high standards and to strengthen the program.
>
> As a dedecated Eagle Scout I urge you to carryon with the mission of NESA, and to strengthen Scouting on an local level. Scouting has enriched our lifes in so many ways, it is our challenge to reach out, and touch the lives of others with a quality Scouting program. Noone has a deeper understanding of the aims and purposes of the Boy Scouts of America than do Eagle Scouts.
>
> Expect nothing from the NESA accept pride in membership the opportunity to serve Scouting and an association with Americas finest citisens. The National Eagle Scout association is designed to give back to Scouting not to continue to receive from it.
>
> Sincere best wishes,
>
> Frederick St. Amour
>
> Frederick St. Amour, Director

After verifying your corrections, choose an adjective to evaluate your proofreading and English skills. ____________________

Practice Quiz

If the sentence is correctly punctuated, write C in the blank; otherwise, write *N* and correct it.

______ 1. The English language is full of traps and pitfalls and it can be harder to write a clear sentence, than to keep a clear conscience.

______ 2. Did he finish the accounting report on Tuesday or, did he attend the controllers' meetings?

*It is correct for an organization to capitalize its "product."

______ 3. First you should type all of today's letters, and then you should file the copies of yesterday's letters.

______ 4. If you wish, he will type all the letters for you and your department.

______ 5. Homemakers who watch TV all day probably serve many TV dinners to their families.

______ 6. Although, he rarely achieves his goals, Mr. O'Connor tries very hard.

______ 7. The diesel's sudden popularity stems from its economy, and its low emissions.

______ 8. Jack Smith and I love the English language more than flowers and wine.

______ 9. Yes, we defend and protect it too much against change.

______ 10. The manager, Amy Lieb, believes that no one is exempt from this ruling.

______ 11. We have mailed you a copy of the first volume, and hope it will reach you this week.

______ 12. The copies dated June 12, 1985 are being sent to you today, not tomorrow.

______ 13. The basic factors for business success are materials, money and management.

______ 14. "In Dayton, Ohio, we have two factories," said Henry Kissinger, Jr. to his father.

______ 15. The convention will begin on March 28, but you don't have to be there, until the next day.

______ 16. The pleasant cheerful young woman knows how to type, file, and take dictation.

______ 17. Please visit us in August if you can't make it in July.

______ 18. Because of a serious misunderstanding we lost the customer, and discharged one of our best salesmen.

______ 19. We need 5000 cheese pizzas for the employees, although I personally prefer pepperoni.

______ 20. Please ship the pizzas by refrigerated air freight to our plants in Akron and in Oakland.

______ 21. Ron, have you met my cousin Chuck?

______ 22. Upon opening the door, Phyllis smelled gas.

______ 23. All the students in the class who had not handed in their term papers failed.

______ 24. "Very well," Steve said "I'll obtain the papers by Friday."

______ 25. In August Ellen plans to reduce the prices on oak furniture.

SECTION 10

PUNCTUATION POTPOURRI

After Completing This Section, You Will:

- *Use the following punctuation with precision: semicolon, colon, period, exclamation mark, question mark, hyphen, dash, comma, quotation marks, apostrophe, parentheses.*

"Rotten pot" is the literal translation of the French word "potpourri." In American usage, however, the word refers to a combination or mixture. On this page the combination or mixture referred to is one made up of useful dots, lines, and curves. Look up "potpourri" in your dictionary to see whether you are pronouncing it correctly; and while you're there, notice its definitions.

Leaving the French now, we move to an Italian of the 15th century named Aldus Manutius, the man chiefly responsible for systematizing punctuation. The principles he developed enabled both writers and readers to interpret a sentence in the same way. Although his system has been modernized through the years and adapted to various languages, its purpose remains the same: to enable a reader to better understand a writer.

One of the best-known examples of double meaning caused by the absence of punctuation is the written reply of the prophets to the Roman soldiers. The soldiers had asked whether they would return from the war. Since death was often the punishment for an inaccurate prediction, the soothsayers understandably took great care with their words: "ibis et redieris non morieris in bello" (you will go and return not die in war).

Because a punctuation system had not yet been devised at that time, the reader could interpret it both ways—with a pause either before or after "not." That way the soothsayers could keep their heads no matter how the war turned out.

If you want to "keep ahead" in business, mastering modern punctuation principles will help you communicate effectively and confidently. Today's punctuation system includes marks that are required in certain positions in writing. Sometimes, however, the placement of a mark is either optional or dependent on the judgment of the writer. The writer's judgment is applied by determining the exact shade of meaning he wishes to put across. For example, the placement of certain punctuation marks can control how important or unimportant an idea looks to the reader. In other cases, omission or inclusion of a particular mark will cause the sentence to be read more smoothly.

When you have a clear understanding of all the punctuation rules *and* you have experience in business writing, you will develop a "sense" of when to apply variations of rules that are optional. In the meantime, this section offers punctuation rules that enable you to punctuate with precision and without risk of criticism. In a few cases, you're offered a choice where you must use your judgment. Most of the rules are, however, sufficiently specific to provide you with confidence in the correctness of your writing.

READ UNIT 58

IN CONCLUSION . ! ?

This first unit of Section Ten begins with the end—the end of sentences. The three marks used to end sentences are the period, the exclamation mark, and the question mark.

USE THE PERIOD

(a) **After a complete thought that is a statement.**

We sent you the bill last week.

(b) **After a complete thought that is a command or a courteous request.**

Pay your bill this week. (command)
Please pay your bill this week. (courteous request)

(c) **After a complete thought that is a courteous request, even though it is worded like a question. If action is desired rather than a reply, use a period. For example, after asking a customer to pay a bill, the seller wants the money, not a "yes" or "no" answer.**

Will you please pay your bill now.
Would you please pay your bill now.

(d) **After a complete thought that is an indirect question.**

I asked whether he would pay his bill.
I wonder whether you will pay your bill.

(e) **After an expression that stands for a complete sentence.**

Yes, of course.

USE A QUESTION MARK

(a) **After a complete thought that is a direct question.**

Will you pay the bill this week or next?
Do you intend to pay this bill?

(b) **After a complete thought that might be considered presumptuous if punctuated as a courteous request.**

Would you please handle my mail while I'm away?
(as part of a memo to your superior in the office)

BUT

Would you please handle my mail while I'm away. (courteous request)
(as part of a memo to your subordinate in the office)

(c) **After a question that is not a complete thought when the rest of the sentence is understood.**

If so, when?
Why?

(d) **In parentheses to express uncertainty.**

Although our new accountant (?) can't count, he is the boss's nephew.

USE AN EXCLAMATION MARK

(a) **To express strong feeling after a complete thought.**

Fill out the card and return it today!
I can't believe it!
Will you please sit down!

The exclamation mark used at the end of what looks like a statement, command, or courteous request enables the reader to sense some strong emotion or urgency. If the expression is spoken instead of written, voice and facial expression transmit the strong feeling to the listener. For example, say these three correctly punctuated sentences aloud:

Will you please go home now?
Will you please go home now.
Will you please go home now!

(b) **To express strong feeling after an expression that stands for a sentence.**

What excellent equipment!
Wait!

Special Notes 1. The exclamation mark is often used in advertising copy and sales letters. It should be used rarely in other types of business writing. Overuse of the exclamation mark causes it to lose its effectiveness—just as the parent who yells at the children all the time finds that the children are no longer affected by the yelling.

2. If your typewriter does not have an exclamation mark, you can type it like this:

   Type a period.
   Backspace once.
   Type an apostrophe.

> Woman! Without her, man would be uncivilized.
> Woman without her man would be uncivilized!
> *Which sentence has the exclamation mark in the proper place?*

REPLAY UNIT 58

Add periods, question marks, and exclamation marks where necessary. (Beware of run-on sentences and comma splices, which are described in Units 11 and 12.)

1. I wonder whether he will be attending the conference
2. You play all day I, however, work hard
3. Would you please ship our order by air express
4. Will you be at the convention
5. Management makes important policies and decisions, I just carry them out
6. He asked me whether I would be at the convention
7. Would Thursday or Friday be more convenient for you
8. Buy UNEEDA now
9. Will you please file these letters
10. Wonderful
11. May I hear from you by return mail
12. Will you come
13. He asked if you would come
14. Your house is on fire
15. What a beautiful day

Check answers on page 325.

READ UNIT 59

THE HALFWAY MARK ;

The semicolon (;) might be called a halfway mark because it is the punctuation midway in "pausing value" between the comma and the period, explaining perhaps why it is made up of one of each.

> Use a semicolon to *join* two complete thoughts that are *not* joined by **and, but, or, nor, for**.

Transition words (such as *however, nevertheless, then,* and *therefore*) often join the two closely connected ideas. Use a semicolon before these transitional expressions when they join **two independent clauses**. Since transition words are usually parenthetical, or nonessential, use a comma after them. However, when the transition is a short word (*then, thus, hence, still, yet, also*), omit the comma.

(a) First she typed the letters; then she made the phone calls.
(b) That typewriter is electronic; this one is not.
(c) He was upset by the criticism; therefore, he refused to discuss the matter.
(d) We're unable to offer him full-time employment; however, we can hire him on a part-time basis.

When you wish to *separate* the two thoughts, use a period. If you're not sure whether to join with a semicolon or separate with a period, use your judgment. Either way will be correct, but one way might read more smoothly than the other.

> Use a semicolon to join two *complete* thoughts that *are* joined by one of these conjunctions—**and, but, or, nor, for**—IF *one or more* commas appear elsewhere in the sentence.

(a) Jo Hansen, who is an expert in this field, will show the African slides; but we expect the others to participate also.
(b) As a matter of fact, his work has been good; and considering his extensive training, we think he will progress rapidly.
(c) We would like to buy that beautiful dog; but the owner, Cathy Allen, will not sell it.

> Use a semicolon after a *complete thought* that comes before **for example** or **namely** or **that is** when the expression precedes a list or an explanation that ends the sentence.

(a) We need some additional equipment to complete the project; for example, several drawing boards and at least two compasses.
(b) Jennifer Crystal Imports, Inc., is conveniently located; that is, just two blocks south of the Long Beach exit of the Long Island Expressway.

> Sometimes a series already has commas *within* the items of the series. In such cases, use semicolons *between* the items.

(a) AMTRAK stops at Schenectady, New York; Springfield, Massachusetts; and West Palm Beach, Florida.

(b) The new officers of the corporation are Mr. David Stern, President; Ms. Mildred Sweet, Vice President; and Mr. Steven Berman, Secretary-Treasurer.

REPLAY UNIT 59

Insert semicolons where needed. If the sentence does not require a semicolon, write *C* in the answer blank.

1. I should like to stay he wants to leave.
2. I should like to stay, but he wants to leave.
3. Our records indicate, Mr. Green, that we filled six orders for you last year and every one of them was delivered within three days.
4. Our records indicate, Mr. Green, that we filled six orders for you last year. Every one of them was delivered within three days.
5. He would like to buy the antique furniture but the owner, David Arant, will not sell it.
6. He has one overpowering ambition namely, to fly a jet plane.
7. Bill has many assets that most people are not aware of, for example, he has a law degree.
8. Carolinda graduated from Shimer College, which is in Illinois therefore, she would be ideal for the job.
9. The president of a big corporation generally earns a higher salary than the President of the United States.
10. A large business is highly complex and difficult to understand it is divided into many departments in which people perform specialized functions.
11. Although as a teacher he doesn't earn much money, he wouldn't trade jobs with anyone.
12. His typing and shorthand speeds are low however, he was hired as secretary because of his excellent English and mathematics skills.
13. This year Standard Packaging might declare a dividend that is, a sum of money paid to stockholders out of the company's earnings.

14. From a study of old newspapers, I find that business conferences were held in this building on March 19, 1902, April 11, 1913, and September 2, 1945.
15. The furniture will be delivered today, but the phones won't be installed until next Monday.
16. I am a great believer in luck and I find that the harder I work, the more I have of it.
17. Although I am a great believer in luck, I find that the harder I work, the more I have of it.
18. I believe, however, that you are right.
19. It isn't hard work that kills, it's worry.
20. The materials arrived late, therefore, the secretary didn't type the report today.

Please check the answers on page 325.

READ UNIT 60

AN EASY MARK :

The rules for the colon (:) are easy to learn:

1. Use a colon after a complete sentence (or independent thought) when a single word follows.

   Just one word describes him: cruel.

2. Use a colon after a complete sentence (or independent thought) when a listing follows.

   The secretary needs these supplies: carbon paper, envelopes, and letterheads.

   BUT

   DON'T use a colon after an *in*complete thought when a listing follows:

   The personality traits that he was most interested in were initiative, loyalty, honesty, and dependability.

   BUT

   DO use the colon after the incomplete thought if the items are listed on separate lines.

   The personality traits that he was most interested in were:

   Initiative
   Loyalty
   Honesty
   Dependability

3. Use a colon after a complete sentence (independent thought) when a second independent thought explains or supplements the first.

   His goal is clear: He wants to be wealthy.
   (When a complete sentence follows a colon, capitalize the first letter.)

4. Use a colon when an independent thought (complete sentence) introduces a quotation or before a long quotation.

   Elwood Chapman gives this explanation: "Although most organizations require many raw materials, machines, equipment, buildings, and much money, a business is made up primarily of people."

5. Use a colon after the salutation of a business letter when "mixed," or "standard," punctuation is used. Standard punctuation means the use of a colon after the salutation as well as a comma after the complimentary close.

   Ladies and Gentlemen:
   Sincerely yours,

   Open punctuation means *no* marks after salutation and complimentary close.

   Ladies and Gentlemen
   Sincerely yours

6. Use a colon between hour and minutes.

   Our appointment is at 12:30 p.m.

> After typing a colon, space twice (except between hours and minutes, as in Rule 6).

REPLAY UNIT 60

Insert colons where needed. Write C in the blank if the sentence is correctly punctuated.

______ 1. Please ship the following two dozen of Style No. 308 and three dozen of Style No. 402.

______ 2. He has just one goal in life revenge.

______ 3. Please send these items to Detroit, Chicago, Seattle, and San Diego.

______ 4. Your most serious problem is money.

______ 5. We plan to visit these cities Detroit, Chicago, Seattle, and San Diego.

6. These are two important things you can do to prevent shoplifting Place mirrors in strategic locations and post special warning signs.

7. Our school is in a lovely community in the San Fernando Valley Woodland Hills.

8. Here is something worth thinking about A small idea that produces something is worth more than a big idea that produces nothing.

9. Judges have a double duty They are supposed to protect the innocent and punish the guilty.

10. On page 156 the author writes "Appearance counts greatly when a person is to be chosen from among a number of people, for appearance unconsciously affects the attitudes of the person who does the choosing."

Insert 2 colons, 1 semicolon, 1 period, 1 question mark, and 1 exclamation mark where needed.

11. Rebecca made an important decision yesterday She decided to hire a new secretary

12. Amy Jaffe, her present secretary, is extremely inefficient but she has a great deal of charm, which is why everyone likes her.

13. The new secretary will need these qualifications 100-word-a-minute dictation speed, 60-word-a-minute typing speed, excellent spelling, and an exacting knowledge of punctuation.

14. Do you think we can find someone with those qualifications

15. What a difficult job for the Personnel Office

Please check your answers on page 325.

READ UNIT 61

GOOD MARKSMANSHIP -- ( )

The business writer who has good marksmanship hits exactly the right spot with even the less frequently used marks—the dash and the parentheses.

Insecure writers (in handwritten work) tend to use the dash like a security blanket. When not sure of whether to use period, comma, semicolon, or colon, they use a dash and hope the reader will assume the writer was in a hurry. Although parentheses errors are rare, some suggestions for appropriate use are included here as a guide.

- Use dashes to set off an abrupt interruption within a sentence.

  My plan saved the company thousands--no, it was nearer tens of thousands--of dollars last year.

- Use dashes to emphasize a parenthetical expression that would ordinarily be enclosed with commas. To de-emphasize such an expression, use parentheses.

  (a) The president of this company, a man who once earned $10 a week as a janitor, is one of the richest men in the world. (ordinary parenthetical expression)
  (b) The president of this company (a man who once earned $10 a week as a janitor) is one of the richest men in the world. (de-emphasis of parenthetical expression)
  (c) The president of this company--a man who once earned $10 a week as a janitor--is one of the richest men in the world. (emphasis of parenthetical expression)

In the sentence above, a good writer would choose the dashes to emphasize an interesting point. However, the commas and parentheses also result in correctly punctuated sentences.

- Use dashes to separate a parenthetical, or nonessential, expression that contains one or more commas.

  The officers of this corporation--the president, the vice president, the treasurer, and the secretary--are in complete agreement about the new budget.

  (Notice how the sentence would be misinterpreted if a comma were used after "corporation.")

- Use a dash after a series that comes *before* a complete thought.

  Dependability, loyalty, and efficiency--those are the qualities we are looking for in an employee.

  BUT

  These are the qualities we are looking for in an employee: dependability, loyalty, and efficiency.

  OR

  The qualities we are looking for in an employee are dependability, loyalty, and efficiency.

- Use parentheses to enclose directions.

  The profits (see chart, page 7) were the highest in the history of the company.

- Use parentheses to set off a supplementary expression.

  Our supervisor (he's a new employee) is an expert in holography.

Note The dash is constructed on the typewriter by typing two hyphens. In handwriting, make it twice as long as a hyphen. The printed dash is also a long line, as after "stone" below.

> Effective use of dash:
> A diamond is the hardest stone—to get.

REPLAY UNIT 61

Insert dashes, parentheses, or colons. Write C beside the correctly punctuated sentence.

1. Harbor Office Supply Company I'll check the address has ordered ten Ace Calculators.
2. This thesaurus a treasury of synonyms, antonyms, parallel words, and related words was first published in 1852 by Peter Mark Roget.
3. These fine machines they are the best money can buy are offered to you for only $99 each.
4. Money, beauty, intelligence, and charm she has all of them.
5. She has money, beauty, intelligence, and charm.
6. She has all these attributes money, beauty, intelligence, and charm.
7. We must see him at once not tomorrow.
8. The job requires the following skills shorthand, filing, typing, and bookkeeping.
9. His check for $152 not $156 was returned by the bank.
10. Roosevelt Island was supposed to be New York City's ideal place to live a crime-free, auto-free, dog-free new island right in the East River.
11. The owner plans to put all the profits for the year $100,000 back into the business.
12. Shorthand, filing, typing, and bookkeeping those are the requirements for the job.
13. The printing on the container was also attractive much more inviting than that on your regular brand.
14. We are not interested not now at least in your proposition.
15. The three departments of our government the executive, the legislative, and the judicial derive their authority from the Constitution.

Answers are on page 325.
The POP QUIZ for Units 58–61 is on page 279.

READ UNIT 62

PLAGIARISM'S ENEMY " "

If you're not sure of what "plagiarism" means, look it up in your dictionary to see why *quotation marks* are its enemy. Whenever repeating someone else's exact words, you are using a *direct quote*. To avoid being accused of plagiarism, use quotation marks when writing a direct quote.

DIRECT QUOTATION

He said, "I don't believe we can proceed in an organized manner."

INDIRECT QUOTATION

He said that he didn't believe we could proceed in an organized manner. (No quotation marks because this is *not* a direct quote; this is a paraphrase—a rewording that gives the meaning in another form. When you paraphrase, you put the statement in your own words.)

DIRECT QUOTATION

Elwood Chapman writes, "A business is an organization that brings capital and labor together in the hope of making a profit for its owner or owners."

INDIRECT QUOTATION

Elwood Chapman writes that the principal aim of a business is to make a profit for its owners. (No quotation marks because this is not a direct quote; it is a paraphrase.)

OTHER USES FOR QUOTATION MARKS (and underlines)

- To show a reader that slang or an informal expression was used for special effect where it might otherwise seem out of place.

  Frankly, I'm too "chicken" to invest in a venture that seems so risky.

- To show that a word or expression is being used to draw attention to that word:

  Many people find it difficult to distinguish between "effect" and "affect."

  Too many "don'ts"* in your life can lead to frustration.
  (The underline may be used for this purpose instead of quotation marks.)

*When forming the plural of a contraction, don't add another apostrophe.

- To enclose titles of subdivisions of published works—such as titles of articles appearing in magazines or titles of chapters. Also use quotation marks for names of films, plays, shows, poems, songs, lectures, and so on.

  Refer to "Notes on Quotes," page 217, to review the use of the comma with quotation marks.

- Full-length published materials (books, magazines, newspapers) should be underlined or typed in all capital letters. In print, italics replace all caps or underlining.

  GETTING INTO BUSINESS by Elwood Chapman is used as a text in college-level introductory business courses.
  (All caps—easier when typing.)

  Getting Into Business by Elwood Chapman is used as a text in college-level introductory business courses.
  (Underline—easier when handwriting.)

MORE NOTES FOR QUOTES

These "notes" review some of Unit 55 as well as introduce additional aids for quotation marks.

- When a period or a comma is needed with the closing quotation mark, *always* (no exceptions) place the quotation mark *after* the period or the comma.

  "Some publishers are born great, some have greatness thrust upon them, and others merely survive television," said John H. Johnson of *Ebony* magazine.

  In 1972 Mr. Johnson accepted the Magazine Publishers Association award for "Publisher of the Year."

- When a colon or a semicolon is needed with the closing quotation mark, *always* place the quotation mark *before* the colon or the semicolon.

  The following scientists were quoted in the article entitled "Rediscovering the Mind": Maureen Conway, Maria Colligan, Carl Sagan, and Georgi Lozanov.

  When we expressed concern, the bookkeeper explained, "The check was accidentally postdated"; however, we have still not received a correctly dated check from him.

- When a question mark or an exclamation mark is needed with the closing quotation mark, determine whether the question or exclamation is part of the quoted words or refers to the rest of the sentence. Then place the **?** or **!** appropriately.

  Did you know that he said, "I love you"? (? refers to rest of sentence.)
  Then he said, "Do you love me?" (? refers to quoted words only.)

REPLAY UNIT 62

A potpourri of punctuation marks is required in these sentences—on your mark, get set, go!

1. He shouted Your house is on fire
2. Your house is on fire he shouted
3. She whispered Are you sure you love me
4. Are you sure you love me she whispered
5. Do you know whether he said I love you
6. After he scolded her he said What a beautiful day
7. Do you know what the phrase negotiable instrument means
8. The following information is quoted from the letter The train leaves Chicago at two oclock and arrives in Boston Massachusetts at midnight
9. Was it Mr. Higgins who said Results are what count
10. Its important that they dont think you are a pushy sales representative
11. Rosemary Wrenn of Long Beach said that she would like the job
12. I believe that Elaine and Michelle should be considered for the openings in the Personnel Department.
13. Alexander Smith who lives in Toronto will be writing the screenplay for us
14. Please tell the chef how many tacos you need for the company party
15. This shipment of stationery the manager clearly stated will arrive on time
16. I have this to say regarding his "abject poverty It is fictitious
17. Kristin said Are you all right
18. M*A*S*H was Ms. Johnsons favorite program
19. Bach Suite No. 2 in B Minor is first on the program
20. Ms. Crock whispered Did you know that the item marked 'fragile' was broken on arrival (Review page 199 for single quote use.)

Please check your answers on page 326.

READ UNIT 63

HALF A DASH -

Half a dash equals one hyphen (-). The hyphen and the dash are two separate and distinct marks of punctuation. Unit 61 includes information on use of the dash. This unit is about the hyphen.

In Unit 22, you reviewed compound nouns and observed that hyphens are used in some of them. Here are suggestions to help you recognize other compound words and determine whether to use a hyphen.

WORDS SPELLED WITH A HYPHEN

Compound numbers from twenty-one to ninety-nine are spelled with a hyphen.

BUT

The numbers *one hundred, five million,* and the like are not hyphenated.

When "self" is used before a *word,* include a hyphen:

self-control, self-respect, and so on

The only exception is *selfsame.* (Of course, *selfish* is not hyphenated because *ish* is not a word.)

When "over" or "under" is used before a word, write the whole word without a space or hyphen:

overpayment, underexposed, and the like

When "re" is used before a word that doesn't begin with **e**, write the whole word without a space or hyphen unless the result might be confused with another word of the same spelling:

rediscover, reheat, re-enter or reenter, re-examine or reexamine

BUT

re-cover (to cover again)

AND

recover
(get better from illness or get something back that had been lost)

When in doubt about other words, consult your dictionary.

Words spelled with hyphens contain hyphens in the entry word. When a dot or an accent mark appears between the syllables, write the word "solid," as one word. Consult your dictionary to see how these entry words appear:

brother-in-law, vice president, inasmuch as, double boiler

The five principles listed and illustrated above refer to words that have hyphens as part of their **spelling**.

Hyphens are also used in a way that cannot be verified in the dictionary; that is, to join the elements of a **compound adjective**. Since the dictionary is of no help in such cases, it is important to understand the principles that follow.

COMPOUND ADJECTIVES

An adjective describes or explains a noun or a pronoun. Sometimes two or three words make up one describing expression. A hyphen or two make such an expression easier to read. Because this kind of adjective consists of more than one word, it is called a compound adjective.

a *brownish-white* object
a *one-story* house
an *up-to-date* report

Notice that if any part of the compound adjective is omitted, either the meaning changes or what remains doesn't make sense.

Use a hyphen (or hyphens) in a compound adjective that comes *before* the noun being described.

1. Buy your *back-to-school* clothes now.
   (Hyphens are used because the compound adjective appears *before* the noun CLOTHES.)

   BUT

   Buy your clothes now for going *back to school.* (no hyphens)

2. The *up-to-date* equipment is in the new classrooms.
   (Hyphens are used because the compound adjective appears *before* the noun EQUIPMENT.)

   BUT

   The equipment in the new classrooms is *up to date.* (no hyphens)

Often, as in No. 1, when the compound adjective is moved away from the noun it describes, the part of speech changes; that is, it is no longer an adjective. Being aware of this grammar change is unimportant. Just remember to hyphenate when the describing expression comes *before* the noun.

If the first word of the compound expression ends with *ly, er,* or *est,* the hyphen is usually not required. If, however, the *ly* word is an adjective, do use the hyphen.

The *fashionably dressed* executive carried a leather bag. (*fashionably* is an adverb.)
My employer is a *friendly*-looking man. (*friendly* is an adjective.)
We found out that she is the *highest paid* executive in Detroit.

WORD DIVISION

The third major use of the hyphen is to divide words at the end of a line. The most important rule is DIVIDE BETWEEN SYLLABLES. When a word is divided elsewhere (fl-ower), the effect of the entire piece of work can be destroyed as the reader wonders, "*Where* did that writer go to school? Or *did* he go to school?" Another general rule for word division is to avoid distracting the reader by a word division that might amuse or momentarily confuse.

Please send me your cat-
alog. Just over the horizon-
tal line, you'll find a number.

} incorrectly divided

Other specific rules are required if your written work is to look professional. These rules are in "Word Division" in the Reference Section at the end of the book. If you keep that page close by (and refer to it) when writing, you will get to know the word division rules so well that before long you'll divide correctly without hesitation.

REPLAY UNIT 63

Part A

Compound adjectives are often used in order to write more concisely. The following sentences are written correctly. However, they could be more concise. Rewrite these sentences so that they include compound adjectives requiring hyphens. Do not change the meaning.

Sample Lionel Barrymore was an actor who was well known.
Lionel Barrymore was a well-known actor.

1. I work in a building that has ten stories.
   I work in a ten-story Building.
2. I need a report that is up to date.
   I need an up-to-date report.
3. My father is a man who works hard.
   my father is a hard-worker.
4. The comment she made is off the record.
   she made an off-the-record comment
5. The case against the company that is based in Dallas was handled in Seattle.
   the case against the Dallas-based company was handled in seattle.

Part B

Some of these sentences contain words that should be spelled with a hyphen or with space between the parts. When in doubt, use your dictionary. Write the correctly spelled word in the blank, or write C in the blank if no correction is needed.

1. The artist feels that recreation of the entire scene is possible. ______
2. What is your favorite form of recreation? ______
3. Our overall objectives are similar, but our methods differ. ______
4. You should report underpayments as well as overpayments. ______
5. My fatherinlaw acts like a commanderinchief. ______
6. He is a selfmade millionaire. ______
7. Do you think our country will ever produce enough oil to be selfsufficient? ______
8. Her goal is to save $5,000 by the time she is twenty-one. ______
9. About one hundred pairs of shoes will go on sale next week. ______
10. A person who acts as though intellect and reason are not very important to solving world problems is often called an antiintellectual ______

Part C

Refer to the word division rules at the back of the book. Use a diagonal (/) to show the correct place to divide these words at the end of a line. IT IS ESSENTIAL THAT YOU USE YOUR DICTIONARY AS WELL AS THE RULES. If a word should not be divided at all, place the diagonal at the end of the word.

function	believe	horizontal	wouldn't
thousands	punctuation	aligned	impossible
interrupt	syllables	stopped	guesswork

Answers are on page 326.

READ UNIT 64

WILD APOSTROPHES

Our punctuation potpourri could not be complete without a reminder about avoiding "wild" apostrophes and using apostrophes exactly where they belong. This unit provides an opportunity to review possessive nouns, which you studied in Section Eight. (The other uses of the apostrophe are not repeated here.)

A possessive noun shows ownership, authorship, place of origin, type of use to which something is put, and time relationship. A possessive noun always ends with an apostrophe and an *s*. These principles determine whether the apostrophe is placed before or after the *s*.

1. To make a singular noun possessive, add *'s*

the boss's office Frank's notebook
Ms. Jones's secretary a week's vacation

2. When the singular proper noun has two or more syllables and ends in an *s* or a *z* sound, you may omit the added *s*, to avoid a word that is hard to pronounce.

Socrates' disciples
Ms. Perkins' report

3. To make a plural noun possessive, first look at the last letter of the **plural** noun.

(a) **If the last letter is** s, **add only an apostrophe.**

the Adamses' factory three weeks' work
ladies' clothes all the attorneys' offices

(b) **If the last letter is not** *s*, **add** *'s*.

the alumni's contributions
men's hats
the children's rooms

DO NOT USE AN APOSTROPHE UNLESS THE POSSESSIVE RELATIONSHIP OF TWO NOUNS IS EVIDENT.

The Joneses own factories all over the world.
JONESES is the plural subject of the sentence. OWN is the verb. No possessive relationship between two nouns is shown.

BUT

The Joneses' factories are all over the world.
JONESES' tells *whose* factories and is therefore possessive. FACTORIES is the plural subject of the sentence.

REPLAY UNIT 64

As you read the daily newspaper, a magazine, or any book, be on the lookout for apostrophes. Cut out or copy (include source) any five sentences that contain apostrophes. Below each, indicate whether the apostrophe is used in accordance with the principles shown in this book and state (in your own words) the rule that was applied.

Sentence 1: ______________________________

____

Is apostrophe used correctly? __________ Rule ______________________________
yes/no

____

____

Sentence 2: ______________________________

____

Is apostrophe used correctly? __________ Rule ______________________________
yes/no

____

____

Sentence 3: ______________________________

____

Is apostrophe used correctly? __________ Rule ______________________________
yes/no

____

____

Sentence 4: ______________________________

____

Is apostrophe used correctly? __________ Rule ______________________________
yes/no

____

____

Sentence 5: ______________________________

____

Is apostrophe used correctly? __________ Rule ______________________________
yes/no

____

____

See POP QUIZ on page 280.

Checkpoint

Whenpeoplestartedwritingtheyputonewordafteranother. Later writers separated words with spaces. Eventually they began to mark their writing with dots, dashes, and curves. Now many writers feel that punctuation is the most important single device that leads to easy reading. Punctuation marks enable a writer to imitate spoken language on paper.

All punctuation marks except for the comma are included in this section. Check off the ones you are confident about using and review the appropriate units for those that seem a bit difficult.

______ Use a period, question mark, or exclamation mark at the ends of sentences. (Unit 58)

______ Use a period after a courteous request, even though it might be worded like a question. (Unit 58)

______ Four reasons are given for correct use of the semicolon. Do you know what they are? (Unit 59)

______ Use a colon after an *independent* thought when: (Unit 60)

(a) a single word follows it.
(b) a list follows it.
(c) another independent thought explains the first thought.
(d) a long quotation follows it.

______ Use a colon after the salutation of a "mixed," or "standard," punctuation letter and between the hour and the minutes. (Unit 60)

______ Use dashes to set off: (Unit 61)

(a) an abrupt interruption.
(b) a parenthetical expression to be emphasized.
(c) a series that already contains commas within the items.
(d) a series that comes *before* a complete thought.

______ Use parentheses to enclose: (Unit 61)

(a) directions.
(b) supplementary expressions.
(c) a parenthetical expression to be de-emphasized.

______ Use quotation marks around direct quotations and around subdivisions of published works. Use the underline or all capital letters to designate titles of full-length materials. (Unit 62)

______ Use quotation marks around certain slang expressions and around an expression used to draw attention to itself. (Unit 62)

______ Three rules determine whether the quotation mark is before or after another mark of punctuation. Do you know them? (Unit 62)

______ The hyphen is used: (Unit 63)

(a) as part of the spelling of a word (in which case, it is shown in the entry word in the dictionary).
(b) in a compound adjective that is placed before the noun being described.
(c) for word division at the end of a line. (Review pages 288–289.)

______ The apostrophe shows ownership, authorship, kind, and time relationship. What rules determine whether to place the apostrophe before or after an *s*? (Unit 64)

______ Unit 48 shows uses for the apostrophe other than in possessive nouns. Take a moment to review these **now**.

> **SIGN IN BOARDING HOUSE**
> Please Clean Tub After Bathing Landlady

Additional punctuation principles are explained on pages 304–307 of the Grammar Supplement in the back of the book.

Special Assignment

Select ten punctuation principles from Sections Nine and Ten. Write each principle as briefly as you can. Then compose a sentence that applies the rule. If you apply more than one principle within the same sentence, you may write fewer sentences. Submit this assignment, preferably typed, to your instructor on ________________.
(date)

Samples

1. *Principle:* Use commas to separate the items of a series.
   Sentence: Alan, Lois, Mimi, and Jack met at 1025 Fifth Avenue.

2. *Principle:* A semicolon may be used to join two complete thoughts that are closely related.
   Sentence: An executive needs to make decisions quickly; in fact, decisions are often necessary before all the facts are available.

Combination method may be used as follows:

1. *Principle:* Use commas to separate the items of a series.
2. *Principle:* A semicolon may be used to join two complete thoughts that are closely related.
   Sentence for 1 & 2: A decision had to be made at once; that's why Mrs. Heffron, Mr. Madden, and Ms. Cates were called in.

Proofreading for Careers

Proofread and correct this article. Look for errors in the application of principles studied in this course so far. For skillful proofreading, use a closed pen to "underline" each line as you read it. Read for content as well as for correctness so that you can spot any kind of error. Use your dictionary and the Reference Manual in the back of this book.

AMERICA'S DEPARTMENT STORES

Door-to door peddling is'nt one of the high status jobs but many famous people got there start there, according to Robert Hendrickson, author of, The Grand Emporiums, a book published by Stein & Day.

Cyrus McCormick, first sold his reapers on the road. Benedict Arnold peddled stockings in the Hudson river Valley before he became this countrys best known traiter. Even the early Rockefellers were roadies of a sort. John D.s father billed hisself as "Dr. William A. Rockefeller the celebrated cancer specialist. Among the very best of the 19th century peddlers was Gimbel, Saks and Field, who eventually opened 3 of the 1st department stores.

Mr. Hendrickson tells us that store-keepers have helped shape our language as well as our economy. Calling dollars bucks is traced to early trading in deerskins. "Getting down to brass tacks comes from country merchants who hammered brass headed tacks into their sales counter's to measure lengths of cloth.

Mr. Woolworth often made unannounced visits to his own stores posing as a customer. He also frequently sent telegrams to each of his stores. One of them read, Good morning. Did you say Good morning to each customer this morning? Frank W. Woolworth.

John Wanamaker opened his first store in Philadelphia in 1861. He set up a huge gong at the front door to welcome each customer but it scarred people out of their minds and he had it removed. Mr. Wanamaker and most other store keepers learned that retailing is a visual eye catching kind of business.

The Neiman-Marcus catalog has included such items as: a 30 thousand dollar solid gold omelet pan, a 3 hundred dollar mink sling for the lady who breakes her arm, and a mouse ranch.

The Grand Emporiums is a lively book and certainly worth reading, however, the author does include some errogenous information: The book states that Arthur Wood a retired chairman of Sears, is the son of Gen. Wood who ran Sears from '28 through '54. While its true that they had the same name and job, they were not in the same family.

The book has 78 pp of photographs and unusually interesting information about department stores, the people who run them and there customers.

After checking your corrected article with a key or with your instructor, answer these questions: What kinds of errors, if any, did you make in completing this proofreading practice (e.g., punctuation, numbers, overlooking errors, abbreviations, etc.)?

How do you evaluate your career English and proofreading skills?

Excellent __ Good __ Fair __ Needs Improvement __ Other? __

Practice Quiz

This quiz includes application of principles from Sections Two, Four, Eight, Nine, and Ten. If the sentence is correctly punctuated, write *C* to the left; otherwise, make the necessary changes.

______ 1. Obviously upset by the criticism he refused to make the necessary changes.

______ 2. Have you read the article entitled "Increase Your Vocabulary"?

______ 3. In the Chicago office, for example, this plan saved the company; thousands of dollars every year for the past five years.

______ 4. His itinerary provides for stopovers in Springfield, Massachusetts, Chicago, Illinois, and Houston, Texas.

______ 5. The personality traits that we were most interested in were; initiative, loyalty, and dependability.

_______ 6. Elizabeth, who is a good listener, never once interrupted while I read what must have seemed to be a never ending story.

_______ 7. When Good Queen Bess, who could swear like a sailor enjoyed one of her rages, the whole Court trembled.

_______ 8. This little booklet contains forty six important facts, and much useful information.

_______ 9. The supervisor said, "Fire him"!

_______ 10. He told us that he had allowed his usual discount; namely, 8 percent off for a cash payment within ten days.

_______ 11. "The sale of mens' and boys' coats will be held next week, said the manager."

_______ 12. Mr. Robertson's automobile was stolen, then his wife ran away with his best friend.

_______ 13. Both events occurred on May 11, 1985 and he wondered, whether this might have been his unlucky day?

_______ 14. Mr. Lewis called Roy and questioned him about the the transaction?

_______ 15. Lily and Penny the two managers of Housewares have requested these items, one additional cash register, a supply kit for the packing table, and two shipping carts.

_______ 16. "The man who lies down on the job," so the lecturer said, "deserves to get run over."

_______ 17. Our enclosed brochure is self explanatory, and may be of interest to you and your employees.

_______ 18. We have sent you a copy of Fundamentals of Business, you may return this textbook at your convenience.

_______ 19. The roads were slippery; therefore, we proceeded with caution.

_______ 20. I warn you I know Im right that the new treasurer is dishonest.

_______ 21. John Wanamaker believed that the American system of storekeeping was the most powerful factor yet discovered to compel minimum prices.

_______ 22. The prices of our greeting cards and desk sets see page 46 of the catalog are subject to a discount.

_______ 23. We hope to hear from you by the first of January; but if we don't, we shall immediately place the account with our attorney's.

_______ 24. Many of today's corporations started in the 1800s.

_______ 25. The competition among business organizations at that time was so fierce that only the strong and aggressive survived.

APPENDIX

FINAL REHEARSAL—SECTIONS 1–10

Correct the errors and write *C* next to the correct sentences. Please do not make a change because you think some other words would "sound" better. Correct specific errors only. Some of the sentences have more than one error; careful proofreading is a must.

1. The passersby throw coins to the musicians who especially pleases them.
2. The sidewalks are an art display in theirselves.
3. Noone is bored when they stroll along the Seine.
4. One of the most spectacular sights are the light demonstration at Notre Dame.
5. This up to the minute information is invalubl e to our company.
6. The architect, as well as the draftsmen, will give their advice on the details.
7. Neither Walter or he wished to give the report to the controller.
8. The secretaries' desks were crowded with yesterdays' work.
9. We've recently learned that Toombs & Irwin, Inc., have been invited to move into the new building.
10. On what criterias did you base your findings?
11. Neither Benedict nor him wished to give the report to the secretary.
12. What is the name of the witness whom we subpoenaed yesterday?
13. I saw the man today whom they say is supposed to be the leader.
14. We carry several kind of three-ring binders.
15. Ability to write shorthand is just one of the skills needed for your success.
16. If anyone calls while I'm out, please ask them to leave a message.
17. There will be a few week's delay in filling your order.
18. The report shows that we are losing money on this product.
19. "An adverb describes an adjective, a verb or another adverb, said the instructor."
20. Neither Ms. Reid nor Lisa are here today.
21. In the files was found a chart and a photograph.
22. The news today are nothing but politics.
23. Each one of us have succeeded.

24. The congregation were quiet.
25. One memoranda was in his box.
26. Either one boy or two girls are going to sing.
27. There is six letters on the desk.
28. Mathematics are an important part of your education.
29. Each committee must make their own rules.
30. The typewriter key was broke when the machine fell to the floor.
31. We hung our coats on the hooks.
32. His sweater has shrank.
33. She give him a book yesterday.
34. The bird has flew away.
35. As soon as the alarm sounded, he sprang into action.
36. He don't intend to visit the showroom today.
37. If I was you, I'd go.
38. I told your manager that you done the work well.
39. He has came to work every day this month.
40. I think he works good under pressure.
41. She cannot afford to buy the mink coat but she might buy a rabbit jacket next week.
42. The defendant has no assets, consequently, there are no prospects of recovering anything.
43. He is lazy, insolent and dishonest.
44. Give the book to John and he.
45. The Crestwood Agency has just completed Marketing Research in the east.
46. The Department of Health, Education, and Welfare are making plans to transfer their headquarters to a larger Building.
47. Two ton trucks are to be driven in the right lane only.
48. Many languages, not only english, is used in the United Nations.
49. We have beat all previous sales records for this territory.
50. No body takes we amateurs seriously.

51. The new manager hired Kathy and I last week.
52. Everyone of the trainess went home.
53. Have you took notes of what the consultant recommended?
54. If you had went with Uriah, you could of helped him.
55. The studio has already chose the actors for the new film.
56. How often has this error occur in the past.
57. Most of the ideas were offered by three people; Rachel, Sean, and Jordan.
58. These three items--the bookcase, the cash register, and the display rack, are not for sale.
59. The owner is real pleased with how you manage the store.
60. The business failed because we did'nt have enough capital.
61. The Floreses are all sopranoes.
62. About one hundred alumnus attended the football game and dance.
63. Please don't buy no more typewriter ribbons until we take inventory.
64. You couldn't hardly expect him to except such a low paying job.
65. If you don't understand the work, you'll get a F on the test.
66. Of the two applicant's we are considering the first seems best qualified.
67. Ms. Mason has done real good in her new position.
68. Our new TV, which works very well, was purchased before the sale.
69. Dame Fortune may smile very sweet today, but tomorrow she may frown real severely.
70. Once Chance has went by there are no future opportunities.
71. Dr. Garvin will be their before dinner.
72. Their planning to expand the New Jersey plant.
73. She will have arrive before the tornado strikes.
74. I be working on this job for a long time.
75. We don't know nothing about those kind of distribution systems.
76. The new policy effected some of we girls in the drafting department.
77. If anyone here wants to object, let them do so now.
78. As she proceeded with her speech, her words flowed more and more easy.

79. They needed to raise the money quick.
80. After we adjusted the carbueretor, the car ran smoother.
81. Yes you may believe me.
82. "Good morning Red Riding Hood" said the big bad wolf.
83. He dug, and hoed his little field, and planted sweet potatos.
84. Innsbruck which is in Austria, is one of the most beautiful citie's in the world.
85. His answers were alright, but his math skill was deplorable.
86. A 16 foot pole will not be long enough.
87. When you take a cruise, you'll find a upper berth is less comfortable than a lower.
88. She is more unhappier than ever.
89. Do you know who's in charge of the barracks.
90. Please remember to destroy them letters.
91. Scotland and Ireland were occupied by Celts, they were known, however, as Gaels.
92. The Romans organized Britain as a province, Christian missionaries converted it.
93. The Celts were known as people who used bronze weapons, but they later learned to use iron.
94. The native food tasted strangely to the American tourists.
95. The general costs of running a business, such as taxes, rent, and heating, is called overhead.
96. A word used to join two words or two groups of words is called a preposition.
97. The derivation or history of a word is known as the usage label.
98. *We, us,* and *ourselves* are usually used as adjectives.
99. Anotnyms are words with similar meanings.
100. If you don't no where your going you'll probably end up somewheres else.
101. It's true that this company's policy provides a discount for payment in thirty days, but your account is eight months over-due.

102. Are all employees' required to use the employee's cafeteria?

103. Milk that has been standing in the hot sun a long time is unfit to drink.

104. The quotations in our price list See page 4 apply to jobbers not consumers.

105. Marks of punctuation act like road signs they tell us when to slow down and when to stop.

106. Mr. Wheeler's 2 daughter-in-laws majored in english literature.

107. Doing things right is not enough, you must do the right things.

108. Just to stand up in the face of lifes problems. That takes courage.

109. Although Philadelphia has been compared to Boston, the resemblance is superficial.

110. A good college or desk dictionary has about a half million entrys.

111. Any insurance man can give you information about rates.

112. She told her that she had made a mistake.

113. Seattle is the nearest of the two cities'.

114. "Why did the peanut cross the road"? asked Ms. Hechanova

115. "To get to the Shell Station"! replied the student.

SPELLING FOR CAREERS

The 3, 2, 1 Plan is an efficient method for mastering the spelling of an entire list of words. Here's how it works:

(3) Write each word in the 35-word group three times, spelling and saying the word aloud (use the dictionary when in doubt about pronunciation).
(2) Next, return to the beginning of the list and write each word twice.
(1) Finally, ask someone to dictate the 35 words to you and then write each word once. With access to a cassette recorder, you can dictate the words to yourself. After checking, list any that you misspelled or felt unsure about; practice these few using the 3, 2, 1 Plan again.

Here are 10 groups of 35 words each. Perhaps you'd like to master one list along with each of the 10 sections of the text.

ONE

abbreviate
absence
absolutely
absurd
abundance
accelerate
accessible
accompanying
accountant
acknowledgment
acquaintance
acquitted
adequacy
adjournment
advantageous
aggressive
align
allotment
allotted
altogether
amateur
ambassador
ambiguous
amendment
amortize
analyze
announcement
annually
annuity
anticipate
anxiety
apologize
apparatus
apparent
appreciable

TWO

approximately
architect
argument
ascertain
assessment
bulletin
buoyancy
bureau
bureaucracy
bylaw
calendar
campaign
canceling
cancellation
candidacy
carriage
category
cautious
census
certain
certificate
challenge
changeable
changing
chargeable
chauffeur
chronicle
colloquial
column
combustion
commencement
compelled
compensate
competent
competition

THREE

concede
conceive
concession
concurred
confidentially
confidently
conjecture
Connecticut
connoisseur
conscientious
conscious
consensus
consistent
continually
controversy
convertible
cooperation
correspondence
corroborate
counselor
courageous
courteous
crisis
criticism
criticize
currency
customary
debatable
deceive
decision
deductible
default
deferred
deficient
definitely

FOUR

deluge
dependent
description
desirable
desperate
determine
deterrent
develop
development
different
diligent
dimension
disappearance
disapprove
disastrous
discrepancy
discretion
disguise
dissatisfied
distributor
divided
dual (double)
duplicate
economical
edition
effervescent
efficiency
eighth
electronic
elementary
eligible
eliminate
emergency
emphasize
encouragement

FIVE

encouraging
endorsement
enforceable
enormous
en route
enthusiastically
entirely
enumerate
environment
equipped
erroneous
especially
etiquette
Europe
exaggerate
excel
excellent
excusable
execute
exercise
exhibit
exorbitant
expedient
expenditure
experience
extension
extraordinary
extravagant
facilities
fallacy
familiarize
fascinate
feasible
fictitious
financier

SIX

fiscal
fluorescent
forcible
foreign
forfeit
forth (forward)
forty
fourth (4th)
fragile
fraudulent
freight
fulfill
fundamentally
generalize
government
grandeur
grateful
grievance
guarantee
handful
harass
hazardous
hesitant
hindrance
hypocrite
immediately
impromptu
improvement
incidentally
indispensable
initiative
insistence
intangible
interpretation
intolerable

SEVEN

irrelevant
itemize
itinerary
jeopardize
jewelry
judgment
justifiable
knowledge
laboratory
launch
league
legible
legitimate
lucrative
maintenance
manageable
management
maneuver
marketable
Massachusetts
measurement
media (plural)
mediator
mediocre
memorize
messenger
miniature
minimize
miscellaneous
necessarily
necessitate
negotiate
nineteen
notarize
numerous

EIGHT

observant
occasionally
occurred
occurrence
omission
outrageous
pamphlet
participant
patronize
per annum
perceive
perceptible
perennial
permitted
petroleum
plausible
politics
possession
precede
precedent
precision
preferable
preferred
prominent
questionnaire
quota
receipt
reconciliation
recurrence
reference
referred
regrettable
reiterate
relevant
remittance

NINE

repetition
representative
rescind
residual
respectability
restaurant
ridiculous
salable
satellite
satisfactorily
scarcity
scheme
scrutinize
seize
sensible
serviceable
similar
solemn
sophisticated
souvenir
specialize
specifically
sphere
spontaneous
statistical
statistician
subsidize
substantial
substantiate
suburb
succeed
summarize
supersede
supplement
surmise

TEN

surplus
survey
synthetic
systematically
taxable
tedious
temperament
tourist
tournament
tragedy
transferred
treacherous
triplicate
ultimately
unanimous
unavoidable
undeniable
undoubtedly
unduly
unprecedented
usable
vacillate
vacuum
validate
vengeance
verbatim
vice versa
visible
whereas
wholly
withhold
workable
writing
yield
zealous

> *From an AP dispatch*—Shanty Hogan, football coach at Phoenix Junior College, claims that this is a true story.
>
> He asked all his freshmen players to fill out a card to be used in case of serious injury. The card lists whom to notify and such information. One blank is for religion. One player wrote, "Bhaptizz."
>
> Hogan chuckled and then asked the lad, "Now, son, what religion are you?"
>
> The boy answered, "Presbyterian."
>
> "But you wrote Baptist," the coach said.
>
> "I know," the player said, "but I can't spell Presbyterian."

SPELLING GUIDES

The guides that follow govern the spelling of many frequently misspelled words. However, use these guides with care because of the many exceptions.

*THE OLD VERSE YOU LEARNED IN ELEMENTARY SCHOOL IS STILL TRUE:

I before E
Except after C
Or when sounded like A
As in neighbor and weigh

For this rule to be most useful, add the idea that the "I before E" guide is applied only to words with the long *e* sound.

- *I* before *E*: achieve, believe, chief, field, fierce, grievance, piece, relieve, siege, wield, yield.
- *E* before *I* after *C*: ceiling, conceit, deceive, perceive, receipt, receive.
- *E* before *I* for long *a* sound: eight, freight, neighbor, reign, surveillance, vein, weigh.

EXCEPTIONS: Neither weird sheik leisurely seized either financier.

*PERHAPS THE MOST USEFUL SPELLING GUIDE CONCERNS ADDING PREFIXES (WORD BEGINNINGS) TO ROOT WORDS:

- When the root word begins with the last letter of the prefix, a double letter results:

mis+spell=misspell	dis+satisfaction=dissatisfaction
im+movable=immovable	il+legal=illegal
un+noticed=unnoticed	ir+responsible=irresponsible

BUT

in+comparable=incomparable	in+animate=inanimate
dis+appear=disappear	re+commend=recommend
dis+appoint=disappoint	un+able=unable

*HERE'S A GUIDE TO WHETHER OR NOT A FINAL SILENT *e* SHOULD BE DROPPED BEFORE ADDING A SUFFIX (WORD ENDING):

- Drop the final *e* before a suffix that begins with a vowel.

enclose	enclosure	advise	advisable
guide	guidance	use	usable
argue	arguing	desire	desirous
arrive	arrival	true	truism

BUT if the word ends with *ce* or *ge*, keep the *e* before suffixes beginning with *a* or *o*

notice	noticeable	advantage	advantageous
enforce	enforceable	service	serviceable

- Keep the final *e* before an ending that begins with a consonant.

encourage	encouragement	manage	management
false	falsehood	sincere	sincerely
hope	hopeful	sure	surely

EXCEPTIONS: acknowledgment, agreeable, argument, dyeing, duly, judgment, mileage, ninth, truly, wholly.

*THE FOLLOWING GUIDES DETERMINE WHETHER TO DOUBLE A FINAL CONSONANT BEFORE ADDING A SUFFIX.

- If the suffix begins with a vowel, double the final consonant of a one-syllable word ending in a single consonant (except *y*, *w*, or *x*) preceded by a single vowel.

plan	planned	run	runner	sad	sadden
ship	shipping	slip	slippage	wrap	wrapping

EXCEPTIONS: bus—buses or busses, busing or bussing, bused or bussed.

- Apply the preceding rule to *two*-syllable words if the second syllable is accented.

admit	admitted	occur	occurrence	refer	referring
confer	conferred	prefer	preferred	transfer	transferring

EXCEPTION: transferable.

- If the accent shifts to the first syllable when you add the suffix beginning with a vowel, don't double the final consonant.

confer	conference	refer	reference	prefer	preferable preference

- When a word of more than one syllable is not accented on the last syllable, don't double the final consonant before adding a suffix.

benefit/benefited	cancel/canceled	credit/crediting
differ/difference	happen/happened	profit/profiting
retail/retailing	total/totaled	travel/traveler

EXCEPTIONS: cancellation, programmed, programming, programmer.

- Regardless of syllables, don't double a final consonant before an ending that begins with a consonant.

commit/commitment	equip/equipment	glad/gladness
hand/handful	sad/sadly	ship/shipment

- Don't double the final consonant when a word doesn't end with a single consonant preceded by a single vowel.

confirm/confirming	exist/existence	look/looking
prevail/prevailing	return/returned	treat/treated

*CHANGE the *y* TO *i*.

- For words ending in *y* preceded by a consonant, change the *y* to *i* before adding an ending.

defy/defiant	happy/happiness	heavy/heaviest
hurry/hurried	likely/likelihood	plenty/plentiful

BUT If the ending begins with *i*, keep the *y*:

try/trying   forty/fortyish   accompany/accompanying

- For words ending in *y* preceded by a vowel, keep the *y* before any ending.

annoy/annoyance   delay/delayed   employ/employable.

EXCEPTIONS: day/daily, pay/paid, slay/slain, lay/laid.

*LOOK UP WORDS ENDING IN *able/ible, ant/ent, ance/ence, ize/ise/yse.*

- No rules govern these suffixes; consult your dictionary.

*IT'S EASY TO SPELL THE SYLLABLE PRONOUNCED "SEED" CORRECTLY.

- *Supersede* is the only word that ends in **sede**.
- *Exceed, proceed,* and *succeed* are the only three words spelled with **ceed**. The spelling remains when adding *ed, ing,* or *s*. However, drop one of the *e*'s when spelling **procedure**.
- All other words with the syllable pronounced "seed" are spelled **cede**: *intercede, precede, recede, secede,* and others.

POP QUIZZES

UNIT NO.'S	PAGE	NO. RIGHT	OUT OF
1–3	263	________	20
4–6	264	________	25
8–10	265	________	25
11–13	266	________	8
15–18	267	________	10
19–21	268	________	30
22–24	269	________	15
25–27	270	________	25
28–30	271	________	20
31–33	272	________	25
34–38	273	________	20
39–42	274	________	15
39–45	275	________	25
46–47	276	________	25
49–52	277	________	25
53–57	278	________	25
58–61	279	________	25
62–64	280	________	25

POP QUIZ—UNITS 1–3, SECTION 1

Please insert nouns in the blanks.

1. The __________________ types all the __________________.
2. In __________________ the __________________ is two feet deep.
3. __________________ wants a new __________________.
4. __________________ is the best __________________.
5. Our __________________ is closed on __________________.

Please insert verbs in the blanks.

6. She __________________ all the way to the office.
7. He __________________ the letter and __________________ it.
8. Ms. Modisette __________________ chairman of the Board of Directors until last year.
9. The students all __________________ military uniforms.
10. The Internal Revenue Service __________________ our tax return.

Please insert pronouns in the blanks.

11. The persons __________________ were trained to use the word processing equipment are now known as correspondence secretaries.
12. Thank __________________ for writing to __________________ about the problem __________________ are having with __________________ copier.
13. __________________ wrote a letter to __________________ about __________________.
14. When __________________ goes wrong with __________________ what will __________________ do?
15. Give __________________ to __________________ will take __________________.

Please insert adjectives in the blanks.

16. Our __________________ copier makes life easy for __________________ operator.
17. Please reserve __________________ rooms for our __________________ members.
18. The vice president presented an __________________ plan for selling your __________________ appliances.
19. The __________________ chairs and the __________________ desks will be used in the attorney's __________________ office.
20. Dr. Kupsh read a __________________ report to __________________ participants.

POP QUIZ—UNITS 4–6, SECTION 1

Please insert prepositions in the blanks and then place parentheses around the prepositional phrases in each sentence.

1. The book is ________________ the desk ________________ the water cooler.
2. New carpeting is needed ________________ our office and ________________ the lobby.
3. ________________ exactly 5 p.m. the mail was delivered ________________ the clerk.
4. Everyone ________________ him walked ________________ the corner.
5. He looks ________________ his brother ________________ head ________________ toe.

Insert conjunctions in the blanks.

6. More ________________ more companies are realizing the advantage of conversion to microfilm.
7. He could qualify for a good job ________________ he knew shorthand.
8. ________________ he knows shorthand, he can qualify for a good job.
9. You may apply for a job as either an administrative support secretary ________________ as a correspondence secretary.
10. Career paths sometimes become limited ________________ supervisory positions may not exist.

Convert these adjectives to adverbs. *Use the dictionary when in doubt about spelling.*

	ADJECTIVE	ADVERB
11.	happy	________________
12.	loud	________________
13.	accidental	________________
14.	true	________________
15.	busy	________________

Insert adverbs that do not end in "ly."

16. He ________________ works during the summer.
17. It costs ________________.
18. He does ________________ need a stapler.
19. She dresses ________________.
20. She is ________________ good at typing tabulations.

What part of speech is the underlined word?

21. Shopping is his favorite activity when he's on vacation. ________________
22. However, his wife prefers swimming. ________________
23. They provided the child with swimming lessons. ________________
24. They are shopping for new equipment for the word processing center. ________
25. They must raise the money before they hire an architect. ________________

POP QUIZ—UNITS 8–10, SECTION 2

______ 1. To be a sentence, a group of words requires (a) identity (b) action (c) independence (d) none of these (e) a, b, and c.

______ 2. The identity requirement of a sentence means that every sentence must have (a) a subject (b) a verb (c) independence (d) a, b, and c.

______ 3. The action requirement of a sentence means that every sentence must have (a) a subject (b) a verb (c) independence (d) none of these.

Write *T* for true or *F* for false.

______ 4. *If, because, since* and *as* are examples of words that make a clause dependent.

______ 5. An independent word group or clause cannot stand alone as a complete sentence.

______ 6. If a dependent word group is joined to an independent, the result may be a complete and correct sentence.

______ 7. A prepositional phrase contains a subject and a verb.

______ 8. A fragment might be an independent word group, or clause.

______ 9. The verb tells what the subject does or is or has.

______ 10. If a word group has a subject and a verb, it is always independent.

Write *F* for fragment or *S* for complete sentence.

______ 11. Among the interests during the Civil War period.

______ 12. As soon as the students arrived and the parents and faculty had gone.

______ 13. Since the establishment of this plan.

______ 14. The architect was seated at her desk.

______ 15. More than any other single segment of society.

______ 16. Industry holds the answers to these questions.

______ 17. Devouring his lunch in a piglike fashion.

______ 18. In a research study funded by the U.S. Office of Education.

______ 19. We are all required to attend the training program.

______ 20. Mr. Sampson, while viewing the slides on the large new screen.

______ 21. The workers, having been on strike for more than three months.

______ 22. Photography, now popular as an art as well as a hobby.

______ 23. Please show Mr. Vergara where the computer is, and then introduce him to the clerks.

______ 24. Ms. Lafitte was seated at her desk.

______ 25. The lecture began after the salespeople arrived.

POP QUIZ—UNITS 11–13, SECTION 2

Write four comma-splice sentences about yourself or members of your family. For example: I attended Theodore Roosevelt High School in the Bronx, my husband went to DeWitt Clinton High School.

. capital
;
, conjunction

1. I went to city college in the summer; My brother went STATE university.

2. I bought a new car this year, last year I bought a motocycle.

3. I went to Santa Cruz Beach Sunday afternoon, Sunday night I went dancing.

4. I come to school 3 day per week, I rest on the weekends.

In Units 11–13, you practiced four methods of correcting a comma splice. One way is to *separate* the independent parts as explained in Read Unit 11. Two ways of *joining* the independent parts are explained in Read Unit 12, and one more *joining* method is presented in Read Unit 13. Please correct the comma splices you wrote above. Use at least three of the four different correction methods.

5. ______________________________

6. ______________________________

7. ______________________________

8. ______________________________

POP QUIZ—UNITS 15–18, SECTION 3

Use your dictionary.

1. What is a synonym for the word RESPONSIBLE? ______________________
2. Show the pronunciation for the city of VALLEJO. ______________________
3. What does it mean if a word doesn't have a usage label? ______________________

   ____
4. What is the etymology of the word SHREWD? ______________________
5. ABSTRACT is used as what parts of speech? ______________________
6. What does the abbreviation AAUW or A.A.U.W. mean? ______________________

   ____
7. Show the first pronunciation given for BONA FIDE. ______________________

   ____
8. What does BONA FIDE mean in relation to an offer to sell or buy something?

   ____

   ____
9. Referring to your dictionary, divide these words into syllables by drawing a diagonal line (/) at the end of each syllable:

   worked beetle getting fullness contagion
10. The treasurer dis__rsed (*pe* or *bu?*) all the funds before declaring bankruptcy.

POP QUIZ—UNITS 19–21, SECTION 4

Spell the plurals of these nouns. Use your dictionary when in doubt.

1. inventory ____________
2. testimony ____________
3. valley ____________
4. motto ____________ or ____________
5. goose ____________
6. louse ____________ or ____________
7. money ____________ or ____________
8. moose ____________
9. hippopotamus ____________ or ____________
10. genius ____________ or ____________
11. Adams ____________
12. Gates ____________
13. stepchild ____________
14. plaintiff ____________
15. bacterium ____________
16. trousseau ____________ or ____________
17. tomato ____________
18. academy ____________
19. knife ____________
20. statistics ____________
21. Chinese ____________
22. ox ____________

Cross out the incorrect words in parentheses

23. Civics (was/were) the most interesting course in the seventh grade.
24. The news (is/are) good today.
25. Shift on the 9 key to type the left (parentheses/parenthesis).
26. The president has to handle several (crisis/crises/criseses) at one time.
27. Your clothes (is/are) all over the place.
28. Even though the factory is on strike, the goods (was/were) shipped yesterday.
29. I typed five (memoranda/memorandas/memorandum) this morning.
30. The president of the Society for Psychic Research introduced two new (medias/media/mediums).

POP QUIZ—UNITS 22–24, SECTION 4

Based on your study of Units 22–24, proofread this page and make the needed corrections. Please use your dictionary.

1. We have two applicants for the Accounting job from Miami Dade Community College.
2. The College is in the south of Florida.
3. The young lady who works for me is studying engineering at night.
4. When the authoress of this book is presiding over a meeting, she prefers to be called the "chairman" rather than the "chairperson."
5. We visited the Empire State building, which is in New York city.
6. We should have a Social committee to plan parties and to buy gifts for special occasions.
7. Last winter we arranged for my brother-in-laws to attend the conference in the Spring.
8. Many passerbys saluted the commander-in-chief.
9. A poetess, a lady doctor, and a woman bank president were among the mothers who agreed to serve as officers of the PTA this year.
10. The attorney general took two spoonsful of honey before he began the cross examination.
11. The tour of the west includes visits to the grand canyon and to Yellowstone national park.
12. Sean Chandra has just been elected to the senate and he will soon see his cousin Rachel, who is a woman senator.
13. What will you do with the proceed from last year's sales?
14. Copys of all his itinerarys are in the portfolioes.
15. A government order prohibiting trade with a particular country is called a tariff.

POP QUIZ—UNITS 25–27, SECTION 5

Please make the necessary corrections. Write *C* next to the three sentences that are already correct.

1. Please tell me when you plan to visit Ms. Rufrano and myself.
2. You're application for credit has been rejected.
3. Do you really think he is better informed on the subject than her.
4. You're taking a leave of absence during the busiest time of the year.
5. The supervisor regrets this incident even more than I.
6. Is Ms. Waxman better qualified than her for the job?
7. No organizations are more respected than them.
8. The ream of stationery on the counter is your's.
9. If you prefer hers, you could request an exchange.
10. The new member should be seated between Ann and I.
11. The engineer thanked the drafting aide and myself for our help with the account.
12. Those blueprints are our's.
13. Him and Ozelle will discuss it at the forum.
14. The owner and me discussed working conditions for the assemblers.
15. Us older employees always try to help the newer workers.
16. You are a better painter than him.
17. After conferring with Henry and they, he left.
18. Us men are out on the field practicing every day.
19. You will hear from either the treasurer or myself within two weeks.
20. Jill and her are the two best soloists.
21. You must remember that our opponents want to win as much as us.
22. Marian married a man who was as critical and intolerant as herself.
23. She herself was the first one to admit it.
24. The coach had a long talk with each of we halfbacks.
25. He asked my brother and I to change the tire for him.

POP QUIZ—UNITS 28–30, SECTION 5

Make the necessary pronoun corrections. If the pronoun is correct, write *C* in the blank. Don't make a change unless the pronoun is INCORRECT.

_______ 1. Someone left his books on my desk.

_______ 2. We may hire a zone manager whom Burroughs Corporation just discharged.

_______ 3. Whom do you suppose will lead the district in sales this month?

_______ 4. Whom do you think we should select for the job of district manager?

_______ 5. It was probably Mr. Robins whom you heard speaking.

_______ 6. The company filed it's suit last week.

_______ 7. Did he say who he wanted to see?

_______ 8. The program was designed by Mr. Rentch, whom we all recognize as a superior systems analyst.

_______ 9. We don't know who is to blame.

_______ 10. Mr. McMullen said everyone of the students did well on the accounting test.

_______ 11. Each girl will be given an application blank, which they will be asked to fill out.

_______ 12. One man after another took his place in line.

_______ 13. Every one of the programmers had his or her own idea on the subject.

_______ 14. We all respect the woman to whom we dedicated this building.

_______ 15. We met the former manager, who had not been in the bank for years.

_______ 16. Every body in this office is being asked to contribute to the fund.

_______ 17. If any customer would like a brochure, they may have one.

_______ 18. The Board of Education moved their offices to Canal Street.

_______ 19. The old company closed their doors yesterday for the last time.

_______ 20. Each employee is to indicate when he wants his vacation.

POP QUIZ—UNITS 31–33, SECTION 6

In the blank at the end of each sentence, write the correct form of the verb given in parentheses. Please use your dictionary as necessary. Spelling counts. Some are correct already.

1. How often has this mistake (occur) in the past year? ______ 1.
2. The mirror was (break) when it was delivered. ______ 2.
3. Many elderly people have (benefit) from her work. ______ 3.
4. Transportation charges on both shipments were (prepay). ______ 4.
5. Mr. Bautista has been (transfer) to our Seattle office. ______ 5.
6. Yesterday Mr. Sean S. Chandra (begin) an in-depth study of the plan. ______ 6.
7. For two years the business has been (lose) money. ______ 7.
8. Do you know who (precede) Marcia Karl in this position? ______ 8.
9. This branch is well (equip) to give good service. ______ 9.
10. Ms. Lindscomb (deserve) praise for teaching all those children to read. (present) ______ 10.
11. Mr. Bond has (lay)* the matter before the board. ______ 11.
12. The balance sheet has been (lie)* here for a week. ______ 12.
13. The weather bureau has (forecast) snow for tomorrow in Milwaukee. ______ 13.
14. How long do you think this carton has (lie)* here? ______ 14.
15. Yesterday Uriah (order) more produce for the Miami store. ______ 15.
16. Our new discount policy (improve) our sales. (past) ______ 16.
17. All our employees (enjoy) the annual barbecue. (present) ______ 17.
18. Last week Howard and Michelle (refer) to increased sales. ______ 18.
19. They now (deny) having made any comments regarding the new bonus plan. ______ 19.
20. The manager is (transfer) most of the new employees to the Anchorage branch. ______ 20.
21. He is, however, (permit) me to remain in Orlando. ______ 21.
22. I am not (imply) that you approve of the plan. ______ 22.
23. The chairman of the book club felt that *How Green Was My Valley* (was) an exciting novel. ______ 23.
24. I'll (defer) shipment of your order until next week. ______ 24.
25. It (cause) his blood pressure to rise. (present) ______ 25.

*Look up the verbs LIE and LAY in the dictionary.

POP QUIZ—UNITS 34–38, SECTION 6

Each of these sentences contains one verb error. Write the correct verb in the blank.

1. Every one of the stenographers do her best.
2. Ms. Dale and Mr. Kennedy is rapid typists.
3. Neither the instructor nor the student are here.
4. Ms. Giron, as well as Mr. Heffron, attend every meeting.
5. Neither the salesman nor the managers was in the office.
6. On my desk is four letters for Mr. Sandell.
7. Half of the envelopes was on my desk.
8. None of the paper were on my desk.
9. Everyone who completes the medical assistant program are qualified to take the examination.
10. The results of this new expression of democracy in education is impressive.
11. Either of these two procedures are practical.
12. Ruth, along with her assistants, work with Jeffrey in the Reprographics Department.
13. A carton of books be there since last week.
14. Congress are for this legislation.
15. I wish that she was more assertive.
16. If he was the manager, he would fire everybody.
17. His educational background, experience, and personality is well suited for the job.
18. A statement of assets and liabilities were prepared yesterday.
19. A list of names and addresses were sent out last week.
20. Although the report were received last week, I've not yet read it.

POP QUIZ—UNITS 39–42, SECTION 7

Choose the correct form of the adjective that is in parentheses. When in doubt, use your dictionary.

_____ 1. Which of the two teams do you think has the (good) chance of winning the World Series?

_____ 2. The fat, poorly dressed man is the (young) of the five cousins.

_____ 3. Two employees were in the same accident, but the older one suffered the (bad) injury.

_____ 4. In the 25 years of our business, she is the (careful) seamstress we have had.

_____ 5. Which is the (much) beautiful city, Dallas or Miami?

_____ 6. As the (heavy) woman on the team, Darla was asked to go on a diet.

_____ 7. Of all the hockey teams in the country, which do you consider to be the (good)?

_____ 8. This is the (bad) display of sportsmanship that I've ever seen.

_____ 9. Of all the cars in the race, the Porsche traveled the (far).

_____ 10. This pen-and-pencil set costs the (little) of all those pictured in the catalog.

Correct the errors.

11. That there lady weighs 450 pounds and it's a honor to know her because she is a star of a circus.
12. Those kind of oranges and them pears would make a good salad.
13. These type of people are a irritation to me when I need a X ray.
14. Which would be more profitable, a $11 increase or a 1 percent increase?
15. Please write a *U* in the first blank and a *S* in the second.

POP QUIZ—UNITS 39–45, SECTION 7

Most of these sentences contain adjective or adverb errors. Show the corrections in the blanks or write C if the sentence is correct. Use the dictionary.

1. Hardly nobody likes that course because it don't offer no challenge.
2. You can't never make no money in those kind of businesses because the overhead is so high.
3. Lindsay Crystal has worked harder in this sales campaign than anybody else.
4. Which of the twins has the best personality?
5. He is angryer than I've ever seen him.
6. Remember that the most safest rule is to drive within the speed limit.
7. He plays the piano good.
8. He is a good piano player.
9. She feels real good today.
10. He seems different today.
11. Smoking is the least of the two evils.
12. The accountant's report is more better than the controller's.
13. She typed the letters as good as possible under the circumstances.
14. His voice sounds too softly without the microphone.
15. When you strike the bar quicker, the machine operates better.
16. The children all sat quiet through the ceremony.
17. She dresses too casual for the office.
18. He was busyier than anyone could have expected.
19. The personnel director will be instructed to be carefuller in her hiring procedures.
20. We needed an one-cent stamp a hour ago.
21. Anywheres you go, you'll find that there kind of person.
22. A apple a day keeps the doctor away.
23. Although she returned back to the hospital two weeks ago, she still feels badly.
24. The food at the new restaurant tasted and smelled badly.
25. While preparing coffee for his boss this morning, he accidentaly overturned the sugar bowl.

POP QUIZ—UNITS 46–47, SECTION 8

Fill in the blanks with the correct forms of the nouns that are in parentheses. One of them is already correct.

1. From Mr. (Sax) letter we learned that prices have advanced 11 percent since Ms. (March) article was published. ____________ ____________
2. If (Schulz) gets the layout into Mr. (Austerlitz) hands today, the advertisement will be run on time. ____________ ____________
3. The signature on the contract isn't my (brother-in-law). ____________
4. Has Mr. (Fingles) committee reached its quota of contributions in the (P.T.A.) annual fund-raising campaign? ____________ ____________
5. During the convention, (Ms. Janowicz) committee will be in charge of registration, and Mr. (Dickens) group will usher. ____________ ____________
6. The (governor) correspondence is being handled by two (secretary). ____________ ____________
7. A successful business is the outcome of meeting every (customer) requirements. ____________
8. The (Commissioner) office issues each (driver) license. ____________ ____________
9. This building was planned for (doctor) and (dentist) offices. ____________ ____________
10. It is our (agency) duty to ascertain all our (customer) desires. ____________ ____________

If the possessive is incorrectly shown, write the correct form in the blank. If the possessive is correct, write C in the blank.

11. President Zeitlin visited Ms. Jackson's business English class. ____________
12. Girl's and women's coats are on sale today. ____________
13. My mother-in-law's will was probated in Tucson. ____________
14. A notaries' seal appears on the bill of sale. ____________
15. Our chairman's decision was omitted from the minutes. ____________
16. There will be a few day's delay in filling orders. ____________
17. After a week's consideration, I've decided to sign the contract. ____________
18. A creditor's meeting will be held on March 2. ____________
19. Passengers' baggage should not be placed in the aisle. ____________
20. I conducted eight salespeoples' conferences last year. ____________
21. The conclusion is the secretarys. ____________
22. The statements do not represent Mr. Jones's beliefs. ____________
23. Beginner's class's start today. ____________
24. They have a large inventory of children's clothing. ____________
25. Addresses of subscribers cannot be changed without two weeks advance notice. ____________

POP QUIZ—UNITS 49–52, SECTION 9

Insert commas where needed by applying the rules in Units 49–52. Write C next to the six sentences that are already correct.

1. In the autocratic style of management the manager draws a very narrow firm discipline line.
2. Before the Hawthorne experiments management felt certain that the way to improve production was to improve the machinery reduce unnecessary hand motions provide better lighting reduce fatigue and so on.
3. As a result of research such as the Hawthorne experiment employers have found that individual recognition opportunity to learn treatment as an individual personal involvement and many other factors are also vital.
4. Personality conflicts are emotionally upsetting and cause serious drops in productivity.
5. Some employees find little satisfaction in their personal lives and have a deep need to enjoy rewarding relationships with those they meet at work.
6. If the foregoing ideas are interesting to you perhaps you should take courses in human relations or in industrial psychology.
7. You might discover that you have a career interest in such a position as personnel director industrial psychologist or human relations consultant.
8. Some important morale factors are satisfaction with the job itself opportunity to learn and compatibility with co-workers.
9. In a corporation having several thousand employees top-level managers seldom know more than a small fraction of those on the payroll.
10. The personnel department is responsible for the activities that concern employees and it advises other departments and management regarding personnel matters.
11. The clerk blushed and handed the message to the tall stern general.
12. Formal courses will not of themselves make you effective but they will help you learn about supervision and management.
13. If becoming a supervisor sounds promising you should start to prepare now.
14. Some decisions may seem trivial to you but the quality of each decision is important to the success of a department.
15. Yes you will need to talk seriously with your people.
16. A supervisor must learn how to conduct group meetings to discipline and terminate employees and to be a good member of the management team.
17. Roberta Simon is the manager of a nationally known variety store that is located in a prosperous upper-middle-class section of Hartford.
18. The store did more than $750,000 in sales last year and is considered successful by the other merchants.
19. Six months ago Esther and Rachel were assigned stores to manage.
20. Yesterday Jordan decided that he wanted to get into management as soon as possible.
21. He may not know everything about how to motivate others but he is willing to learn.
22. The supervisor must learn early in her career that what others do under her guidance is more important than the actual work she does herself.

23. Should you move into a supervisory job from a technical one you must quickly become a people-oriented individual.
24. Whenever you can take at least a few management courses.
25. If you are interested in becoming a manager maintain a courageous open-minded attitude toward your problems.

POP QUIZ—UNITS 53–57, SECTION 9

Applying the rules from Units 53–57, insert commas in these sentences. Write C next to the sentences that don't require additional commas.

1. We noticed that Ivy has changed her methods not her objectives.
2. The typing test was given to all the applicants; the shorthand test to a selected group.
3. Payment for half the purchase price $250,000 is due on June 1; the balance on October 15.
4. We need an electronic calculator, not a 10-key adding machine.
5. Your report was completed on June 3; Dawn's, June 16.
6. The name of the book is *How to Catch a Wolf*.
7. On January 3 1988 we shall meet in Longmeadow Massachusetts to discuss the marketing plan for the new products.
8. Please send the scripts to my nephew Alex who lives in Toronto with his parents.
9. You will need two new writers for the offices in Vance, Canada and in Calgary, Canada.
10. I believe Phil that a more experienced person is needed for the job.
11. The president Edward Robings Ph.D formerly taught business English and communication.
12. They live at 40536 Picket Fence Road Levittown Pennsylvania in an older home that is in very good condition.
13. The textbook used this semester *Word Processing: Concepts and Careers* will inform you of new career paths and how to train for them.
14. It was the expense of the work not its difficulty that made us refuse to begin the project.
15. Wrestling not modern dance was the physical education class I wanted.
16. "You must study during the winter vacation," stressed Mr. Worley "because final examinations begin in January."
17. Please telephone Janice and Billy when the costumes are ready.
18. Ms. Green who is Mr. Williams' administrative assistant is able to take dictation at 120 words a minute and is a CPA.
19. They replied "We think we are within our rights."
20. "Do you think so?" she asked.
21. All birds, especially those that migrate fascinate me.
22. It is I'm pleased to report, a promising idea.
23. I never never eat tomatoes although my son eats them often.
24. I tell you Mr. Marino there is no other way to settle this case.
25. The real reason however was quite different from the reasons he gave.

POP QUIZ—UNITS 58–61, SECTION 10

( ) . ? ! -- ; : Parentheses, periods, question marks, exclamation marks, dashes, semicolons, colons

Please insert only the punctuation marks shown above. I've put the commas in already. Write C beside the two sentences that are already correct.

1. Sales last year were well over ten million dollars the best year in our history.
2. A lawn mower, a rake, and a garden hose these are the implements I must buy next.
3. In the United States men own, on the average, three coats an overcoat, a topcoat, and a raincoat and five pairs of trousers.
4. May I hear from you by return mail
5. In the course of the tour, we shall visit four points of interest in Washington

   Library of Congress
   White House
   Capital
   Smithsonian Institute

6. Everything seems favorable for increased production The workmen are experienced, the morale is excellent, and the machinery is new.
7. Campers must take these items pillows, blankets, linens, and cooking utensils.
8. Yesterday we sent you this night letter "Make shipment to our Harrisburg factory."
9. Our prices are reasonable for example, $9.50 for a ream of heavy-weight typing paper and $4.42 for 500 envelopes.
10. Get a better job then you will earn more money.
11. There are three requirements for the job speed, accuracy, and intelligence.
12. The colors I like best are blue, green, purple, and yellow.
13. Consumer demand can be changed or influenced in other words, you can be influenced to select one item over another.
14. The judges had rejected all but three of the dogs a cocker spaniel, a boxer, and an Irish setter.
15. His phenomenal success is due to one thing hard work.
16. I am planning to visit Oakland, California Homewood, Illinois and Omaha, Nebraska.
17. My assistant and I wish to stay however, Mr. Allen feels we should leave.
18. Will you please answer this letter immediately.
19. Do your records indicate, Mr. Green, that we filled six orders for you last year and that every one of them was delivered within five days
20. He has an overpowering ambition namely, to fly a jet plane.
21. The mayor said, "Our city is growing steadily every year as a result of the expansion of our industrial plants and the friendly attitude shown to visitors by the citizens of our city."
22. I will ask him by telephone assuming he has a telephone.
23. How dare he do such a thing
24. I just cannot believe it
25. These unlawful acts for that is what they are should be forbidden.

POP QUIZ—UNITS 62–64, SECTION 10

" - : ; , '

Please insert only the following in the sentences below: **quotation marks, hyphens, colons, semicolons, commas,** and **apostrophes.** Just one sentence requires no additional punctuation.

1. Supervisors make decisions on an hour to hour basis.
2. A good supervisor is a people oriented person as well as a job oriented person for he or she must get the job done through others.
3. "If you can't take the heat get out of the kitchen" said Harry S. Truman.
4. After eating the elephants yawned and then dozed.
5. As you know nothing was done about the shortage.
6. He likes to eat in first class restaurants in far flung places.
7. All even numbered districts will have elections on Tuesday November 2, 1989.
8. He needs a 12 inch ruler and a two ton truck.
9. On any floor where we did manufacturing such as the second floor the rental was $5 a square foot.
10. We are operating a railroad not conducting a brokers office.
11. Did you read the book PEOPLE ARE FUNNY?
12. I read the article entitled People Are Funny.
13. At first said Ms. Nim the capacity of our plant was small the efficiency low and the performance of equipment uncertain.
14. The wholesale salesperson serves a limited and select class of buyers the retail salesperson meets a large number of people every day.
15. All my products have a money back guarantee.
16. The attorneys office is in that building but she's on a weeks vacation right now.
17. An advertising agency provides a client with

    Research and marketing strategy
    Market forecasts
    Analysis of media possibilities
    Execution of media contracts

18. Insurance companies are divided into two major classifications those that deal with property and liability and those that deal primarily with life insurance.
19. Shirley could remember only three dates 1492 the year Columbus discovered America 1948 the year of her birth and 1963 the year President Kennedy was assassinated.
20. Did Nancy say that she hoped never to set foot in Idahos fields again?
21. That scientific book is out of date for example it contains nothing about lasers.
22. Felixs job is in the mens department.

Use a diagonal (/) to show where these words may be divided at the end of a line. If there is no approved division, place the line at the end. Use your dictionary if necessary.

23. preferred   obliged
24. stopped   capital
25. permitting   selling

READ AND REPLAY REFERENCE MANUAL

READ ABOUT NUMBERS IN BUSINESS

Numbers are an important part of business and technical writing. They are written in figures on invoices, orders, requisitions, and so on, as well as in statistical writing. This page, however, is useful as a reference for determining whether to spell out a number or write it in figures in a letter or a report. The rules given below are a consensus of the style used by better business writers today.

1. General

(a) Spell out numbers up to ten; use figures for specific numbers over ten. When numbers over ten are used in a related way in the same category, however, use figures for all.

We need five electronic engineers in our Dallas plant.
We need 5 electronic engineers, 25 typists, and 30 assemblers in our Dallas plant.

(b) Spell out approximate numbers that can be written in one or two words.

Nearly five thousand employees were laid off last year.
We have developed about a hundred new by-products during the past year.

(c) When a number begins a sentence, spell it out unless it requires more than two words. In that case, rephrase the sentence.

Six hundred crates were shipped to you yesterday.
Yesterday 642 crates were shipped to you.

(d) When expressing millions or billions, make the reading easier by combining figures with words.

We produced 1½ million electric fans during the past fiscal year.

Our company's gross sales during that same period were 66 million dollars.

(e) When two numbers appear together, spell out the number that can be written with fewer letters.

This architect has already designed twenty 16-unit apartment buildings.
This architect has already designed 25 sixteen-unit apartment buildings.

2. Time

(a) Use figures with a.m. and p.m. (Lowercase letters are preferred.) Notice that the colon and zeros are not used for "on the hour" time.

The conference will be held from 9 a.m. through 5:30 p.m.

(b) When not using a.m. or p.m., either spell out the time or use figures.

The conference will begin at nine. (or nine o'clock, nine in the morning, 9 o'clock, etc.)

3. Dates

(a) Use figures when the date follows the name of the month; when the date is in this position, never use *th, d, st, rd,* or *nd.*

The American Bankers Association will meet on May 7 this year.

(b) In military, foreign, and some government correspondence, the date usually appears this way:

The American Bankers Association will meet on 7 May this year. (or 7 May 198__)

(c) In ordinary business correspondence, *of* may be placed between date and name of month if desired. In that case, use *th, nd, rd,* or *st* after the figure, or else spell out the number.

The American Bankers Association will meet on the 27th of May.
The American Bankers Association will meet on the seventh of May.

(d) When the date is mentioned without the name of the month, follow rule (c).

The American Bankers Association will meet on the 27th.

(e) Spell out or use figures for centuries and decades

This book is about nineteenth-century poets. (or 19th-century)
He was a college student during the sixties. (or 1960s)
During the early 1900s many immigrants who had been victims of cruel persecution in their native lands arrived in the Unites States.

4. Money

(a) Use figures for specific amounts of money. Notice that decimal point and zeroes are not used unless for the purpose of being consistent with other amounts in the same context.

A $5 registration fee is required for membership.
Members pay a $5.00 registration fee and $5.50 a month thereafter.

(b) Approximate amounts of money are spelled out.

We'll need about fifty thousand dollars for remodeling.

(c) Spell out the word *cents,* but use figures for the number.

The plugs cost 8 cents each.

(d) Use the dollar sign and decimal point style for *cents* when necessary for the purpose of being consistent with other amounts used in the same context.

The project requires 1,000 wheat chiz at $.15 each, 42 sedus at $7.80, and one large ramir at $82.00.

(e) In legal documents the amounts are usually spelled out (notice capital letters) and then written in figures, which are enclosed in parentheses. In ordinary correspondence, do not repeat numbers.

The fee for use of said property is to be Two Hundred and Fifty Dollars ($250.00) a month and is to be paid on the first. . . .

5. Addresses

(a) Spell out names of streets under 11th.

> The store is on Sixth Street.
> Randy Wahl moved to 11th Avenue in Newport.

(b) Use figures for all house numbers except One.

> Their new suite is at One Abercrombie Street.
> The Atlanta Factory is at 8 Leland Avenue.

6. Percentages and Fractions

(a) Spell out *percent*, but use figures for the number.

> Unemployment was at the 5 percent level that year.

(b) Spell out a common fraction when there is just one in the sentence.
Use figures for less common fractions.
Use figures for a mixed number (fraction and whole number).

> We have received only one-fourth of our order.
> The new specifications require 3/8 of an inch.
> Our profits are 4½ times those of last year.

7. Measurements

Spell out measuring words, such as feet, pounds, and inches but use figures for the numbers.

> The boards are 5 by 6 by 2 inches. (Use "by" instead of "x".)
> Each one weighs 6 pounds 4 ounces.
> The mine is 8 miles away.

8. Age

Usually spell out age that is expressed in years only, unless it immediately follows the person's name. Notice that no commas are used in the third example, which illustrates age expressed in years as well as months and days.

> Mr. Weber will be sixty-three on the day of the presentation.
> Carl Weber, 63, is the new chairman.
> The birth records show that she is 24 years 5 months 6 days as of this date.

9. Books

Use figures for numbers of pages, chapters, volumes, and the like.

> The information you need is on page 46.

REPLAY NUMBERS IN BUSINESS

Change the number style where necessary so that the following sentences would be correct in the paragraphs of a business letter or a formal business report.

1. At 9 o'clock on June 4th, we will have completed forty-two jobs.

2. There are about 50 different ways to make 50 million dollars.

3. 21 boxes were shipped by Acme Freight on 6 June 198-.

4. Please send $5.00 for the book and fifty cents for postage.

5. On page six his age is given as 40.

6. Almost 5,000 attended the Alliance for Survival rally.

7. Our new office is at 62 4th Street.

8. The prime interest rate went to eleven % today.

9. We need 12 8×10 offices in the new building for the members of the sales staff.

10. This store opened its doors at eight a.m. on the 31st of June.

READ ABOUT CAPITALIZATION*

Capitalization is the process of giving a word special importance or emphasis. Hence the first word of a sentence is capitalized to give it the emphasis a word in this important position deserves. Many specific things have two names: the classification name, such as *girl*, and the official name, such as *Mary*. Some of the conventions of capitalization in business writing are explained in Section Four. This page, however, serves as an easy reference for other capitalization questions.

1. General

(a) Capitalize the official names of specific people, animals, places, days, months, holidays, gods, documents, and historical events.

Joseph, United Airlines, Uganda, Veteran's Day, Declaration of Independence, Fabulous Forties, Atomic Age, Wednesday, March, and so on.

(b) Do not capitalize seasons of the year.

fall, winter, spring, summer

(c) Capitalize titles, headings, and the first word of each item in an outline. Use lowercase for short prepositions, "to" as part of an infinitive, articles, and the conjunctions *and* or *or* unless one of these words begins the line.

Titles or Headings: How to Cook With Electricity
Outlines: I. How to cook with electricity

2. Titles of People

(a) Capitalize a title that directly precedes a person's name.

Professor Washington, Reverend Juan Perez, Colonel Modisette

(b) Do not capitalize the title when the name of the person or the title are parenthetical and require commas.

Our English professor, David Stern,. . . The captain, Patrick O'Connor, was . . .

*See Read and Replay 22 for instruction in capitalization.

(c) Do not capitalize the title when it appears *after* the name of the person, unless it is a title of very high government or church rank.

> Michelle Wrenn, mayor of Long Beach, . . .
> Elizabeth, Queen of England, . . .
> The Senator from Florida is the president of Elks Club International.

(d) Do not capitalize names of occupations, such as typist, lawyer, accountant, engineer, and the like.

3. Titles of Publications and Art Works

Capitalize the first word and all principal words of published works, including movies, plays, songs, and so on. Do not capitalize articles (*a, an, the*) or prepositions with fewer than four letters unless they begin the title. Titles of full-length published works should be either underlined or typed in all capital letters (not both). Titles of portions of full-length works (articles, chapters, etc.) as well as movies, plays, poems, songs, and so on should be in quotation marks.

> I read GONE WITH THE WIND (or Gone With the Wind).
> Section Eight is entitled "The Taming of the Apostrophe."
> I saw "Gone With the Wind" at the Bijou.

4. Names of Organizations

(a) Capitalize names of organizations and of specific government groups.

> Supreme Court, Sheimer College, Royal Inn, Department of Internal Revenue, United States Army, Greenland Paper Company.

(b) Unless you are preparing a legal document or a very formal communication, do not capitalize words such as *company, department, college*, and so on when they are used without the name.

> Dawn attends college in Washington.
> Our company will not issue any more common stock this year.
> Give the papers to the chairman of the committee.

(c) Capitalize names of departments within your own organization but not in other organizations.

> Our Shipping Department has packed your order.
> Does your purchasing department have our latest catalog?

5. Names of Places

(a) Capitalize complete names of specific places.

> Atlantic Ocean, Victory Boulevard, Yosemite National Park, Mississippi River

(b) When two or more specific places are named, do not capitalize the plural noun that completes the meaning.

> the Atlantic and Pacific oceans; the Missouri and Mississippi rivers

6. Compass Points

(a) Capitalize compass points that name areas thought of as geographical, cultural, or political units.

Far East, West Coast, Midwest

(b) Do not capitalize compass points that indicate direction or name general areas.

The sun sets in the west.
Drive north along Main Street.
He would like to settle in northern Massachusetts.

7. Trade Names

Capitalize the trade name of a product but not the product word itself unless you work for the company.

Chef Boyardee Pizza (if you work for the producer)
Nescafe coffee (if you don't work for the producer)

8. Business Letters

(a) Capitalize all titles used in inside address or signature. (Do not use Mr. or any other courtesy title in a signature. Always use a courtesy title in an address.)

Mr. Morris Garber, Assistant Manager
or
Mr. Morris Garber
Assistant Manager

(b) Capitalize first word of complimentary close.

Sincerely yours

(c) Capitalize first word of salutation and any noun or title in salutation.

Dear Friends
My dear Mr. President

9. Family Relationships

(a) Capitalize family relationship title when used as part of the name or instead of the name.

Do you think Uncle George will retire soon?
Do you think Uncle will retire soon?

(b) Do not capitalize family relationship title when a possessive noun or pronoun comes before it.

Steve an Sue's uncle is the principal of Saturn Street School.
I believe my cousin should apply for the job.

10. School Subjects and Degrees

(a) Capitalize official names of classes. Do not capitalize the name of a subject or course when it is not the official name—except for languages, which are always capitalized.

Mr. Ivener will teach Business 38.
Mr. Sandell has taught business law for the past three years.
Are you planning to take Spanish or business English this year?

(b) Capitalize the name of a degree when it is used as a title directly after the person's name. Do not capitalize degrees used in any other way. Notice the comma before and after the degree when the degree follows a name.

Margie Sorenson, Ed. D., teaches at Golden West College.
Clara Chung is about to receive a bachelor of science degree.

11. Government Terms

Do not capitalize governmental terms that are used instead of the full official name.

This state has the highest income tax rate in the nation.
We don't wish to violate the city ordinance.

Ethnic Terms

(a) Nationalities and religions are capitalized.

Dutch, Korean, British, Catholic, Hindu, Jewish.

(b) Races are not capitalized when named by color but are capitalized when designated by sociological terms.

black, white, yellow, red and so on, but Negro, Caucasian, Oriental, Indian

REPLAY CAPITALIZATION

Correct the capitalization so that the following sentences will be correct in a business letter or a formal report.

1. The atomic age may be said to have begun on August 5, 1945, when the atomic bomb was dropped on Hiroshima.
2. I flew on American airlines last Summer with the prime minister of Israel and the senator from south Dakota.
3. Eric A. Smith is an independent Auditor who conducts audits for various Department Stores.
4. Winston Churchill wrote Triumph and Tragedy, an important book about world war II.
5. The supreme court decision was favorable to my Company.
6. By september I will know how to speak spanish well enough to take a College course in spanish 2.
7. The Atlantic and Pacific Oceans are natural borders of the United States.
8. I prefer Hunt's tomato sauce for Spaghetti, but my Aunt Mimi says that it contains too much salt.

9. Lalitha, a Hindu woman from the South of India, has a bachelor's degree in ed-

   ucation.

10. The salutation of a letter: Dear customer:

    The complimentary close: Sincerely Yours,

READ ABOUT WORD DIVISION

The picture-frame effect results in an attractive-looking business letter. Left and right margins should be approximately equal. In order to maintain a neat right margin, it is sometimes necessary to divide a word at the end of a line. Because a divided word tends to distract the reader, certain word division conventions have been established to promote greater readability.

I. A careful writer never divides:

   1. the last word on a page.
   2. a word containing an apostrophe.
   3. a number expressed in figures.
   4. an abbreviation.
   5. on more than two consecutive lines.
   6. a word that has fewer than five letters.
   7. a word that has just one syllable.
   8. a proper noun.

II. A careful writer may divide:

   1. between syllables. When consulting the dictionary, refer to the syllables in the entry word, not in the pronunciation. Dots, spaces, and accent marks indicate the syllables.

      fol•low•ing—may be divided between the *l*'s or after the *w*.
      stopped—may not be divided because it is all one syllable.

   2. when at least two letters of the word can be typed on the line before the hyphen.

      OK re-veal NO a-gainst

   3. when at least three letters of the word can be typed on the next line.

      OK compil-ing NO compa-ny

   If the syllables do not permit the minimums of rules 2 and 3, do not divide the word at all.

   4. after a one-letter syllable, unless the vowel is part of a word ending such as *ible* or *able*.

      OK cata-log OK credit-able, incred-ible

   5. between double consonants unless the division breaks the spelling of a root word

      OK begin-ning OK spell-ing

   If the second of the double consonants is needed to form a suffix, place the hyphen between the double letters.

OK posses-sion NO possess-ion

6. at the hyphen when that mark is part of the spelling of the word.

OK self-confidence NO self-confi-dence

7. nonhyphenated compound words between the main parts of the word.

OK under-developed

8. between two vowels that have separate sounds.

OK cre-ation NO crea-tion

III. A careful writer may separate:

1. a date between the day and the year.

Harold and Esther were married on March 8,
1976, in Albany.

2. the first name or middle initial from the last name.

The most capable auditor we have ever had is Mr. Martin L.
Simon.
The most talented dancer in the troupe is Abby
Asher.

3. a long title from the rest of the name.

The most outstanding teacher in our English Department is Professor
David Stern.

4. an address between the street name and the city or between the city and the state.

We have moved our office to 990 Kennedy Street,
Minneapolis, Minnesota.
We have moved our office to 990 Kennedy Street, Minneapolis,
Minnesota.

REPLAY WORD DIVISION

Write *T* (true) or *F* (false) in the blank.

_______ 1. It's OK to divide a four letter word if it has two syllables.

_______ 2. Never divide the last word on a page.

_______ 3. It's OK to divide a one-syllable word as long as it has more than five letters.

_______ 4. Never divide a number expressed in figures; for example, 1,500, -463.

_______ 5. Never divide a word containing an apostrophe; for example have-n't.

_______ 6. If you can't carry at least three letters to the next line, don't divide the word.

_______ 7. If it's necessary to divide a date, type the month on one line and the day on the next. (April—next line, 18)

_______ 8. If an address must be divided, separate the house number from the name of the street; for example, 2166—next line, Clinton Avenue, Bronx

_______ 9. A hyphenated word may be divided only at the hyphen; for example, self-propelled.

_______ 10. Divide between double consonants unless the division breaks the spelling of a root word; for example, spell-ing but win-ning.

READ ABOUT ABBREVIATIONS

Spelling words out instead of abbreviating them has a favorable psychological effect on the reader. It helps to create an image for the company of thoroughness, carefulness, and accuracy. Also, it does not take much more time to spell out a word than to abbreviate it. Thus better writers avoid most abbreviations in letters and reports unless they are technical or statistical. Abbreviations, however, very commonly appear in many other kinds of business communications, such as interoffice memos, informal notes, invoices, receipts, and all kinds of business forms, as well as statistical and technical material.

I. Items A–H below are the abbreviations that *are* acceptable in business correspondence.

(a) Mr., Mrs., Dr., Jr., Sr., Esq., Ph.D., CPA, and so on (*Ms.* is correctly typed with a period, even though it isn't an abbreviation of anything.)

Ronald Rosenberg, Esq. (In the U.S., *Esq.* is used only after an attorney's name.)

(b) Military and professional titles before a name may be abbreviated if used with the full name of the person. When in doubt, spell it out, except for *doctor. Dr.* is preferred with either full name or last name only.

The Rev. Jonathan Flaherty or The Reverend Jonathan Flaherty
but
The Reverend Flaherty
Dr. Melanie Sweet or Dr. Sweet

(c) Names of well-known organizations when you are certain the reader will recognize the abbreviation. Notice that the periods are often omitted in capital-letter abbreviations.

CIA, FBI, AFL-CIO, CBS

(d) Time words.

a.m., p.m., EST (Eastern Standard Time), B.C., A.D.

(e) Parts of names and addresses that are officially abbreviated, such as *Inc., Co.,* or *Ltd.* (Limited) when they appear that way in the company letterhead, or *St.* (Saint) or *Mt.* (Mount) when this is how the place name is spelled (St. Louis). Spell out words such as *street, avenue,* or *boulevard,* and names of cities. (See part (h) for state abbreviations.)

(f) Common business abbreviations and acronyms when you're sure the reader will understand them.

AT&T—American Telephone and Telegraph
c.o.d. or COD—collect on delivery
e.o.m. or EOM—end of month
etc.—and so on
Ext.—when followed by a number for a telephone extension (Ext. 302)
FIFO—first in first out
f.o.b. or FOB—free on board (the point from which the customer pays shipping charges)
GM—General Motors
GNP—gross national product
IBM—International Business Machines
Inc., Corp., or Ltd.—when part of a company's name; Incorporated, Corporation, Limited
IRS—Internal Revenue Service
LIFO—last in, first out
Memo—memorandum
Messrs.—plural of Mister (abbreviation of the French "Messieurs")
NASA—National Aeronautics and Space Administration
No.—before a serial number (Style No. 2348)
P.O. Box—Post Office Box No. 0000
PR—public relations
PS or P.S.—postscript
R & D—research and development
RSVP—please respond (translated from French)
SEC—Securities and Exchange Commission
TV—television
VIP—very important person

(g) An academic degree is abbreviated after a person's name. When it is not used after a name, abbreviate only if you're sure the reader can interpret the abbreviation. Most of these abbreviations are used either with or without the periods.

A.A.—Associate in Arts
A.A.S.—Associate in Applied Science
B.A. or A.B.—Bachelor of Arts
B.B.A.—Bachelor of Business Administration
B.S.—Bachelor of Science
D.A.—Doctor of Arts
D.B.A.—Doctor of Business Administration
D.D.—Doctor of Divinity
D.D.S.—Doctor of Dental Surgery or of Dental Science
Ed.D.—Doctor of Education
Ed.M.—Master of Education
J.D.—Doctor of Jurisprudence
J.M.—Master of Jurisprudence
J.S.D.—Doctor of the Science of Laws
LL.B—Bachelor of Laws
M.A. or A.M.—Master of Arts
M.B.A.—Master of Business Administration
M.D.—Doctor of Medicine
M.S.—Master of Science
Ph.D—Doctor of Philosophy
Th.D—Doctor of Theology

(h) State names appearing in the address may be abbreviated if the official Post Office abbreviation is used. The ZIP follows the abbreviation.

Alabama	AL	Montana	MT
Alaska	AK	Nebraska	NE
Arizona	AZ	Nevada	NV
Arkansas	AR	New Hampshire	NH
California	CA	New Jersey	NJ
Colorado	CO	New Mexico	NM
Connecticut	CT	New York	NY
Delaware	DE	North Carolina	NC
District of Columbia	DC	North Dakota	ND
Florida	FL	Ohio	OH
Georgia	GA	Oklahoma	OK
Hawaii	HI	Oregon	OR
Idaho	ID	Pennsylvania	PA
Illinois	IL	Puerto Rico	PR
Indiana	IN	Rhode Island	RI
Iowa	IA	South Carolina	SC
Kansas	KS	South Dakota	SD
Kentucky	KY	Tennessee	TN
Louisiana	LA	Texas	TX
Maine	ME	Utah	UT
Maryland	MD	Vermont	VT
Massachusetts	MA	Virginia	VA
Michigan	MI	Washington	WA
Minnesota	MN	West Virginia	WV
Mississippi	MS	Wisconsin	WI
Missouri	MO	Wyoming	WY

Spell out the name of the state if you don't use the official abbreviation as shown above. Two-letter Post Office abbreviation should be used *only* when followed by a ZIP code.

II. Here are some other abbreviations often used in business communications of various kinds. These would, however, usually be spelled out in the body of a business letter or in a formal report.

acct.	account
amt.	amount
assn. or assoc.	association
bal.	balance
b.l., b/l, or B/L	bill of lading
c/o	care of
ctn.	carton
cwt.	hundredweight
dept.	department
e.g.	for example
et al. (used after names of people)	and others
frt.	freight
ft.	foot, feet
g.	gram
gal.	gallon
h.p.	horsepower
i.e.	that is
lb.	pound
LCL	less than carload lot
mdse.	merchandise
mfr.	manufacturer
mgr.	manager
misc.	miscellaneous
mo.	month

p.	page
pp.	pages
pd.	paid
rec'd or recd.	received
retd. or ret'd	returned
Sec. or Sec'y	secretary
supt.	superintendent
viz	namely
vs.	versus
yd.	yard
yr.	year

REPLAY ABBREVIATIONS

True/False

_______ 1. To save time in business communications, abbreviate as much as possible.

_______ 2. A good writer will not abbreviate in statistical and technical material.

_______ 3. It's important to include the periods in abbreviations of names of well-known organizations.

_______ 4. It's correct to abbreviate an academic degree when it appears after the person's name.

_______ 5. When typing an address in a business letter, spell out the word "Boulevard."

_______ 6. When "Saint" is part of the name of a city, always spell it out.

_______ 7. Miss., Okla., Calif., and Mass. are examples of correct abbreviations to use in the address of a letter.

_______ 8. A military title may be abbreviated when it precedes the full name of the person.

_______ 9. When writing a business letter, it's correct to abbreviate business terms such as account (acct.) and amount (amt.).

10. What do the folowing abbreviations stand for?

LIFO ______________________ SEC ______________________

MBA ______________________ EOM ______________________

GRAMMAR SUPPLEMENT FOR THE EXPERT

The following pages summarize grammar terminology excluded from various sections. If you learned traditional grammar in the past, you will find this supplement a helpful review. When referring to a Supplement section, review the main section as well. Most of the main section information is not repeated here.

SECTION 1—PARTS OF SPEECH

Traditional grammar recognizes eight parts of speech, seven of which are defined in Section One. (The eighth is the interjection, a single exclamatory word such as *Oh, Gosh*, and so on.) A word, a phrase, or even an entire clause is designated as a particular part of speech by identifying its function in a sentence.

I. Nouns and pronouns function as subjects, objects, or subject complements; any word that functions in one of those roles is automatically a noun or a pronoun.

A. Subjects
Locating a subject is explained in Sections 2 and 7. Although subjects usually appear at the beginning of sentences, they sometimes appear elsewhere.

Although never a mother, she loved children. (subject—she)

B. Objects

1. *Direct Object of a Verb*
Existence verbs do not have objects. To find out whether an action verb has an object, say the subject and the verb; then follow with the question *Whom or What?* If there is an answer, that answer is the direct object of the verb. A verb used with a direct object is known as a transitive verb:

The secretary is typing a letter.

(subject) (verb)
secretary is typing *whom or what?*—answer is *letter.* Therefore, *letter* is the direct object of the verb *is typing*. When there is no answer to *Whom or What?* the verb is known as intransitive:

The secretary is typing quickly.

(subject) (verb)
secretary is typing *whom or what?—quickly* doesn't answer either of those questions; therefore the verb *is typing* does not have a direct object and is intransitive.

2. *Indirect Object of a Verb*
Only a verb that has a direct object may also have an indirect object. To locate an indirect object, say the subject, verb, direct object; than ask *To whom?* or *For whom?* If there is an answer, that answer is the indirect object:

We gave the president a gift.

(subject)	(verb)	(direct object)	
We	gave	gift	*To whom?* or *For whom?*—answer is *president.*

Therefore, *president* is the indirect object. You can tell that *president* isn't the direct object because we can't *give* the president—but we *can* give something *to* him.

3. *Object of a Preposition*
A noun or pronoun that follows a preposition is the *object* of that preposition. Examples are shown in Read and Replay Units 5, 25, and 74. Each object (like all other words) serves just one function in a sentence. Therefore, if a word is an object of a preposition, it cannot be an object of a verb or a subject:

The elephant in the zoo ate two bags of peanuts.

(subject)	(verb)	(direct object)	(objects of prepositions)
elephant	ate	bags	zoo, peanuts

C. Complements
As stated in B1 above, existence verbs do not have objects. A noun or pronoun that follows an existence verb is a subject complement. It isn't necessary to ask any "whom" or "what" questions. A *complement* is something that *completes* or makes whole. A subject complement completes the meaning of the subject by renaming it in some way (noun or pronoun) or by modifying it (adjective).

Susan's mother was the president. Susan's mother is attractive.

(subject)	(existence verb)	noun (complement)	(subject)	(existence verb)	adjective (complement)
mother	was	president	mother	is	attractive

When an existence verb (*was*) links a subject (*mother*) to its complement (*president*), the verb is known as a linking verb. If this subject complement is a noun or pronoun, it is usually possible to reverse the positions of subject and subject complement.

He is my father. My father is he.

II. Conjunctions function as joining words. Subordinate conjunctions (discussed below) sometimes *begin* sentences, however, because the clauses are reversed instead of being in "natural" order.

A. Coordinate Conjunctions
These are used to join parts of sentences that are grammatically equal. In addition to the five included in Unit 5, *so* and *yet* are sometimes used this way. The use of "so" as a coordinate conjunction is not recommended for formal writing.

B. Subordinate Conjunctions
These are used to introduce adverbial clauses, which are always dependent. Some words often used as subordinate conjunctions are *although, when, if, after, unless, even though, as, since, because, so that, whether, before, provided, while*. When they are used as subordinate conjunctions, they precede a subject and a verb. The result is an adverbial clause.

Gail and Eric will take their vacations together *if* they can.

If the order of the clauses is transposed, the subordinate conjunction begins the sentence:

If they can, Gail and Eric will take their vacations together.

Some of the words shown above as subordinate conjunctions are often used as prepositions, in which case they will be the first word of a *phrase*, not a clause:

Prepositional Phrases—after midnight, *since* five o'clock, *until* Christmas

A number of these words may also be used as adverbs, in which case they will neither have an object nor introduce a clause:

We have seen these paintings *before*.

C. Correlative Conjunctions
These come in pairs:

either . . . or   not only . . . but also   neither . . . nor   both . . . and

D. Conjunctive Adverbs
These are transition words that bridge the ideas from one *independent* clause to another. They function as joining words (conjunctions) as well as modifying words (adverbs). Here are some samples: however, nevertheless, moreover, thus, then, otherwise, accordingly, consequently, therefore, furthermore, yet, additionally.

Generally speaking, a contract is valid whether it is written or oral; *however*, certain cases require specific forms.

When the words listed above are *not* used to join independent clauses, they are usually functioning as adverbs:

The courts have ruled, *however*, that an advertisement to sell a car is not a genuine offer.

III. Word groups (phrases and clauses) may function as nouns, adjectives, and adverbs:

(a) He said that you would go. THAT YOU WOULD GO is the direct object of the verb SAID; therefore, the entire dependent clause is a *noun clause*.
(b) She is the manager who makes the most trouble. WHO MAKES THE MOST TROUBLE modifies (describes) the noun MANAGER; therefore, the entire clause is an *adjective clause*.
(c) To know me is to love me. TO KNOW ME is a *noun phrase* because it is the subject of the sentence. TO LOVE ME is also a *noun phrase* because it is the *subject complement* following the *linking verb* IS.
(d) We'll ship the five dozen sweaters today if they're in stock. IF THEY'RE IN STOCK is an *adverb clause* because it modifies the *verb* SHIP.

SECTION 2—SENTENCES

I. The Parts

A sentence must contain a subject and a predicate.

A. Subject—The simple subject is a noun or pronoun that tells who or what the sentence is about. The complete subject is made up of the simple subject plus any words or phrases that modify it. A compound subject consists of two or more subjects joined together—such as "John and Mary."

B. Predicate—The simple predicate is the verb. The complete predicate is made up of a verb plus any modifiers or complements (completing expressions). A compound predicate consists of two or more verbs joined together—such as "work and play."

C. **Clause**—A group of related words containing a subject and a verb (predicate) is a clause.
   1. Independent Clause—A clause that is grammatically complete in itself. It can stand alone as a sentence.
   2. Dependent Clause (also called a subordinate clause)—A clause that begins with a subordinate conjunction. It cannot stand alone.

II. The Sentence

A sentence must consist of at least one independent clause.

A. **Simple Sentence**—This is one independent clause not joined to any other clause: I love you.

B. **Compound Sentence**—This consists of two or more independent clauses joined to make one sentence: I love you and I always will.

C. **Complex Sentence**—This consists of one independent clause plus one or more dependent clauses joined to make one sentence: I love you more than words can tell.

D. **Compound-Complex Sentence**—This is made of two or more independent clauses plus one or more dependent clauses. (It is a compound sentence and complex sentence combined into one): I love you and I hope that you love me.

E. **Fragment**—This is a related group of words (phrase or dependent clause) mistakenly used instead of a sentence: More than words can tell.

F. **Run-on Sentence**—This is a sentence formed by two or more independent clauses joined without a conjunction: I love you I always will. (wrong)

G. **Comma Splice**—When a comma (without a coordinating conjunction) is placed between the clauses of a run-on sentence, the result is called a comma splice: I love you, I always will. (wrong)

III. Constructing Sentences to Avoid Run-Ons and Comma Splices

A. **Compound Sentences**—two or more independent clauses.

   1. Join the independent clauses with a comma plus a coordinating conjunction (and, but, or, nor, for).

      The president of the jewelry company comes from Hungary, but he is now an American citizen.

   2. Join the independent clauses with a semicolon only.

      The president of the jewelry company comes from Hungary; he is now an American citizen.

   3. Join the independent clauses with a semicolon plus a conjunctive adverb (however, nevertheless, then, etc.) or other transitional expression (for example, in fact, etc.).

      The president of the jewelry company comes from Hungary; however, he is now an American citizen.

B. **Simple Sentences**—one independent clause.
Separate the independent clauses into two simple sentences.

The president of the jewelry company comes from Hungary. He is now an American citizen.

C. Complex Sentences—one independent clause and one or more dependent clauses.
Subordinate one of the independent clauses, and join it to the remaining independent clause.

Although the president of the jewelry company comes from Hungary, he is now an American citizen.

SECTION 5—PRONOUNS

I. Case

Although both nouns and pronouns may be classified by "case," identification of case is not important with nouns. However, understanding case of personal pronouns and who/whom is essential to the person who wants to be certain about using the correct form according to traditional grammar.

A. Nominative Case Pronouns

1st person	I, we
2nd person	you
3rd person	he, she, it, they
who	

Use nominative case pronouns for the following:

1. *Subjects*

   Terri Brickman and he plan to go together to the conference.
   Who may I tell her called? (WHO is subject of CALLED.)

2. *Subject Complements* (after linking verb)

   It was probably *she* who asked for the new typewriter.
   It might have been *they*.

B. Objective Case Pronouns

1st person	me, us
2nd person	you
3rd person	him, her, it, them
whom	

Use objective case pronouns for the following:

1. *Direct Object of Verb*

   Please try to motivate Debbie and *him* to do the work.
   Whom shall I take to the concert?

2. *Indirect Object of Verb*

   Please give the president and *me* all the papers on the Dickinson case.

3. *Object of Preposition*

   Please give all the papers on the Dickinson case to the president and *me*.

4. *Subject of an Infinitive* (Infinitives are defined in Section 6 of Supplement.)

I want Myrna and *him* to go.

C. Possessive Case Pronouns

1st person	my, mine, our, ours
2nd person	your, yours
3rd person	his, her, hers, its, their, theirs
whose	

In addition to uses explained in Section 5, use a possessive case pronoun to modify a verbal noun (also known as a *gerund*). Gerunds are defined in Section 6 of Supplement.

DO	I don't approve of *his* working overtime.
DON'T	I don't approve of *him* working overtime.

II. Antecedents

Every pronoun has an antecedent. The antecedent is the noun or pronoun that means the same as the pronoun.

John ate his sandwich. (*John* is the noun antecedent for *his*.)
Everyone did *his* own work. (*Everyone* is the pronoun antecedent for *his*.)

The pronoun must agree with the antecedent in number (singular or plural) and gender (masculine, feminine, or neuter).
If *nor* or *or* joins the elements of the antecedent, the pronoun should agree in number with the antecedent that comes *after* the joining word.

Either the president or the *vice president* will see you in *her* office.
Either the president or the *managers* will see you in *their* office.

When a singular and a plural antecedent are joined by *or* or *nor*, use the plural one after the joining word. (See previous sentence about vice president and managers for example of singular and plural antecedents.)

SECTION 6—VERBS

I. Tense

A. The three simple tenses are past, present, and future. These are explained within the main section on verbs.

B. The three perfect tenses are past perfect, present perfect, and future perfect. The perfect tenses are formed by combining *have, has,* or *had* with a past participle. They enable the writer or speaker to express the time of the action more precisely than can be done with the simple tenses.

1. *Past Perfect*=had+past participle of main verb.
   The past perfect tense shows that the action was completed at a specific time in the past.

   I *had planned* to visit the Ford plant while I was in Detroit.

2. *Present Perfect*=has or have+past participle of main verb.
   The present perfect tense shows that the action began in the past but

was completed just before the present, is being completed in the present, or is continuing.

She *has worked* on that report for six hours.
I *have known* him for a long time.

3. *Future Perfect*=will have or shall have+past participle of main verb. The future perfect tense shows that the action will have been completed before some specific future time.

   When we get married next June, we *will have known* each other for five years.

C. All six tenses mentioned in A and B above may be used in the progressive form. A progressive verb is the combination of an appropriate form of TO BE and the present participle of a verb. It shows that action is in progress.

1. *Past Progressive*=was or were+present participle.

   He *was working* hard all day.

2. *Past Perfect Progressive*=had been+present participle.

   He *had been working* hard all day.

3. *Present Progressive*=is, am, or are+present participle.

   He *is working* hard today.

4. *Present Perfect Progressive*=have been or has been+present participle.

   They *have been working* hard all week.

5. *Future Progressive*=will be or shall be+present participle.

   They *will be working* hard next week.

6. *Future Perfect Progressive*=will have been or shall have been+present participle.

   They *will have been working* hard for two weeks before their vacation begins.

D. Only the past and present tenses have special forms for the *emphatic*. As the name implies, this form shows emphasis. An emphatic verb is the combination of *do* with the basic form of the verb.

1. *Past Tense*=did+basic form.

   Don and Susan *did live* in New York in 1975.

2. *Present Tense*=do or does+basic form.

   Evelyn and Stan *do live* in Santa Monica.
   William *does* go to law school.

II. Shall/Will—Should/Would

A. Traditional Use of Shall/Will

1. For expressing simple future, use *shall* with first person singular or plural pronouns. Use *will* with all other pronouns and with all nouns.

I *shall prepare* a report for your advertising department.
We *shall be pleased* to assist you.
They *will ship* your order early next week.
The auditors *will examine* the records next month.

2. For expressing strong determination or firm promise, use the reverse of the simple future forms. Use *will* with first person singular or plural pronouns and *shall* for all others.

   I *will* finish college no matter what happens.
   These rules *shall* be enforced strictly.

3. *Shall* is used with first person pronoun subjects for asking questions about the future.

   Shall I prepare a report for your advertising department?

B. Colloquial Use of Shall/Will
Will is generally used with all noun or pronoun subjects for statements about the future.

I *will* prepare a report for your advertising department. We *will* prepare a report.

C. Traditional Use of Would/Should

1. For expressing future action, use *should* or *would* in the same way as *shall/will*. See above, A1 and 2.

   I should be pleased to assist you in preparing this report.
   He *would* like to help you also.
   We *should* like to see you at our showroom again.

2. *Should* is used with all subjects to mean "ought to."

   She should get to work on time.

3. *Would* is used with all subjects to mean something that habitually occurs.

   She would get to work on time every morning.

D. Colloquial Use of Would/Should
Would (or will) is generally used with all noun or pronoun subjects for statements about the future.

I *would* be pleased to assist you in preparing this report.

E. Contractions
Contractions using a pronoun with *would* or *should* are acceptable for oral communications (I'd, we'd) but should be avoided in written business communication. However, contractions using a first or second person pronoun with *will* or *shall* are acceptable for nearly all forms of business communication. (The only exceptions would be an unusually formal report or speech and a legal document.)

I'll = I will or I shall
we'll, you'll = we will or shall, you will

III. Subjunctive Mood

The use of *were* to express subjunctive mood is explained in Unit 38. An additional use of the subjunctive requires the use of *be* after such verbs as *order, insist, ask, request, demand,* and *move* (for motions at meetings).

I *move* that the meeting *be* adjourned.
Mrs. Alston *demands* that she *be* fired.
We *ask* that the work *be* done by noon.

IV. Verbals

Verbals are forms of verbs that do not function as verbs. They are used instead as nouns, adjectives, or adverbs. The three kinds of verbals are gerunds (also called verbal nouns), participles, and infinitive.

A. Gerunds or Verbal Nouns

As the alternate name implies, a gerund is a word that looks like a verb but always functions as a noun. Gerunds always end in *ing*.

He is majoring in marketing. (*Marketing*, a gerund, is the object of the preposition *in*.)

A gerund functions as a noun in all ways except that it resembles a verb because it may have an object.

Marketing ice to Eskimos would be rather difficult. (*Marketing* is the gerund subject of the verb *would be*, but *ice* is the object of the gerund.)

B. Participles

A participle is a word that looks like a verb but functions as an adjective.

1. The present participle is used to show that the time is the same as that of the verb in the sentence.

   The woman *wearing* the red dress is a reporter. (*Wearing* is a participle used to introduce a "participial phrase" *wearing the red dress*. This phrase modifies the noun *woman*. The verb is *is*.)

2. The past participle shows that the action took place in the past.

   The merchandise *ordered* last month will be shipped tomorrow. (*Ordered*, the participle, begins the participial phrase *ordered last month*. This phrase modifies the noun *merchandise*, which is the subject of the verb *will be shipped*.)

3. The perfect participle also indicates action that took place in the past.

   Having ordered the merchandise last month, they are eager to receive it. (*Having ordered* is the participle modifying the pronoun *they*.)

   Although participles are adjectives that look like verbs, they may have objects. *Merchandise* is the object of the participle *having ordered*.

C. Infinitives

An infinitive looks like a verb but functions as an adjective, an adverb, or a noun. It consists of *to* plus the basic form or the past participle of the verb. Sometimes *to* is omitted but understood.

We plan *to dictate* the reports now. (present tense)

We expected *to have dictated* them yesterday. (perfect tense)

In the two sentences above, the infinitives are nouns because they function as the object of the verbs *expected* and *plan*.

Careful writers avoid splitting infinitives with modifying words when the split is not necessary for clearness or smoothness.

AVOID	He wants *to* quickly *do* the work. (Adverb *quickly* splits the infinitive.)
USE	He wants *to do* the work quickly. (Move the modifier to a logical place in the sentence.)

V. Transitive and Intransitive Verbs

1. A transitive verb is an action verb that has a direct object. A direct object answers the questions *Whom?* or *What?*

   tr v dir obj
   Steve *teaches industrial arts* in Saratoga Springs.

   A transitive verb may also have an indirect object. After saying the verb and direct object, ask *To whom?* or *For whom?* If there is an answer, that answer is the indirect object. A prepositional phrase may replace an indirect object.

   Fran *gave Jenny* a new *toy*.
   Fran gave a new toy *to Jenny*. (prepositional phrase)

2. An intransitive verb either is a state-of-being verb (existence verb), which never has an object, or is an action verb that does not have an object.

   Fran and Fred *live* in Laurelton.

VI. Agreement of Subject and Verb

All, any, none, more, most, some or a fraction may be either a singular or plural subject, depending on the concept being referred to. If a prepositional phrase follows one of these subjects, the object of the preposition will be the word that determines whether the subject should be singular or plural.

SINGULAR	*All* of the money *is* going to Valerie McDonald.
SINGULAR	*None* of the pie *was* eaten.
PLURAL	*Half* of the boys *are* going on the trip.
PLURAL	*Some* of the storekeepers *have* continued to receive the bulletin.

SECTION 7—ADJECTIVES AND ADVERBS

I. Comparison of Adjectives

A. Absolute Adjectives
Careful writers do not compare adjectives that represent an "absolute" idea. For example, either a room is *square* or it isn't. How can one room be "squarer" than another? Since *unique* means "one of a kind," how can one restaurant be "more unique" than all the others? Here are some other adjectives that should not be used in comparative or superlative forms:

complete	dead	genuine	perfect	stationary	wrong
correct	empty	parallel	round	unanimous	right

A substitute for comparative and superlative forms is available for the types of words shown above: *more nearly* or *most nearly* used with some absolute adjectives enables one to make a logical kind of comparison.

His work is *more nearly correct* than yours.
This carton is the *most nearly empty* of all.

B. Else/Other

When using the comparative form to compare one noun or pronoun with all the others in its classification, exclude the original person or object from the classification. Use *other* or *else* to "exclude" and thus achieve a logical comparison.

ILLOGICAL	He types better than anyone in the office.
LOGICAL	He types better than anyone *else* in the office.
ILLOGICAL	The Denova text editor is better than any I've seen.
LOGICAL	The Denova text editor is better than any *other* I've seen.

II. Adjectives with Existence Verbs

Although the use of adjectives after existence verbs is mentioned in the main section, the explanation that follows might clarify the rule for those who understand grammar terminology:

When a linking verb joins a subject to a modifier, that modifier should be an adjective. Two names are used for this adjective: *predicate adjective* and *subject complement.* Hence we see that a subject complement will be a noun, a pronoun in the nominative case, or an adjective.

The work is correct.
(IS=linking verb; CORRECT=adjective complement)
Ms. Van Meter seems happy today.
(SEEMS=linking verb; HAPPY=adjective complement)

III. Possessive Adjectives

Possessive nouns and pronouns are often called adjectives because they really function as such. A possessive modifies the noun or pronoun that it possesses in much the same way as an adjective. Study these expressions:

green pencil *my* pencil *boy's* pencil

The adjective *green,* the possessive pronoun *my,* and the possessive noun *boy's* all modify the noun *pencil.*

SECTION 8—APOSTROPHES

Here are some additional principles for guidance in the use of possessives.

1. To show that something is owned by more than one person (joint ownership), make the final noun or pronoun possessive. To show separate ownership, make each noun or pronoun possessive.

   Rozini and Marino's factories are now in seven cities.

   BUT if they dissolve their partnership and divide the factories between them,

   Rozini's and Marino's factories are in competition with each other.

2. If a title comes after the name of the "possessor," it is better to use the prepositional phrase instead of the possessive.

DON'T Sidney Berger, Ph.D.'s, report on Chaucer and Keats, won all the literary awards.
DO The report on Chaucer and Keats, written by Sidney Berger, Ph.D., won all the literary awards.

3. Use a possessive form to modify a verbal noun. A verbal noun is a word that looks like a verb because it ends in *ing*, but it is functioning as a noun. (Remember that a part of speech is determined only by how the word functions in the sentence. More on verbal nouns appears in Section 6 of Supplement.)

   The doctor appreciated *Ann Blank's training* of the new assistant.
   The manager was greatly impressed by *Mr. Powell's working* overtime to finish the accounts.

4. Add *'s* after an abbreviation of a singular noun that is possessive.

   Nancy Cohen & Co.'s catalog arrived today.

5. Add only an apostrophe if the abbreviation is plural possessive.

   Two M.D.s' offices are on this floor.
   or
   The offices of two M.D.s are on this floor. (not possessive)

6. Avoid using the possessive form for inanimate objects when it results in an awkward-sounding construction.

   DON'T The chair's leg is broken. (awkward)
   DO The leg of the chair is broken

   BUT

   DO Old Man Winter's coat is covered with ashes. (personification)
   DO The critic tried to explain the portrait's appeal. (not awkward)

7. When an identifying expression follows the name of the owner, avoid using the possessive form; instead, rephrase the sentence.

   DON'T Gary Goetz, our chief engineer's, new office is in Hartford.
   DON'T Gary Goetz's, our chief engineer, new office is in Hartford.
   DO The new office of Gary Goetz, our chief engineer, is in Hartford.

SECTION 9—COMMAS

Below are traditional grammar terms for punctuation principles explained in Units 50–53.

Unit 50

Use a comma between **coordinate adjectives** when *and* is omitted.

A manager needs an open-minded, adaptable attitude for making sound decisions.

Unit 51

Use a comma before a **coordinating conjunction** that joins the independent clauses of a compound sentence.

Chain organizations have regional and district offices, and individual store units are supervised by these offices.

Unit 52

Use a comma after an **introductory verbal phrase** unless the verbal phrase functions as the subject of the sentence.

NO COMMA *To talk with the president* is an honor.
USE COMMA Although not responsible for the general training of salespeople, buyers instruct salespeople on the merchandise they are to sell.

Use a comma after an **introductory adverbial clause**.

Although the buyers are not responsible for the general training of salespeople, they instruct the salespeople on the merchandise they are to sell.

Use a comma after an **introductory prepositional phrase** that has five or more words.

In addition to buying merchandise for sale, buyers are also in charge of sales in their departments.

Use a comma after an introductory prepositional phrase that has fewer than five words when the comma is necessary to avoid confusion.

In addition to selling, salespeople help maintain stock.

Unit 53

Set off with commas **parenthetical expressions**.

Last week, for example, five people were absent because of illness.

Set off with commas **nonrestrictive phrases** and **clauses**.

In the small enterprise, such as a dry-cleaning business or a shoe store, all employees may report directly to the owner or manager.
Local managers of chain stores are permitted to buy goods needed in the store from local dealers, particularly when immediate delivery is desired.

Set off with commas **nonrestrictive appositives**.

At the head of the merchandising division is Ellen D. Alderman, the general merchandise manager.
Ellen D. Alderman, the general merchandise manager, is at the head of the merchandising division.

SECTION 10—MISCELLANEOUS PUNCTUATION

Units 58 and 62 (periods and quotation marks)

Ellipses show that words have been left out of quoted material. Three periods (. . .) with spaces between them show that words have been omitted. Four periods (. . . .) show that the omitted words ended the sentence.

According to the newsletter *Printout*, ". . . Continental Group, the largest U.S. can manufacturer, saved over $60,000 on type costs during 1982, ..."

Unit 63

Hyphens in **compound adjectives** are explained in the main section. Note that if the first word of the **modifier** is an adverb ending in *ly*, the hyphen is not used.

> The *newly developed* procedure is not yet available for general use.

When the compound adjective includes an interrupting word, follow the form shown below:

> Organizing the fall merchandise is usually a one- to two-day job.
> Several two- and three-ton trucks are usually parked outside the diner.

Extra

1. Brackets [ ] enclose a parenthetical expression that is within material already in parentheses. They also enclose extra words inserted in quoted material.

   The student wrote, "The personal [sic] division does not actually supervise store workers."

2. Single quotes, also known as inside quotes, enclose a quotation within another quotation.

   Faña Spielberg, the keynote speaker, said, "Nineteenth century writer William Wallace alluded to the power of women when he wrote, 'The hand that rocks the cradle is the hand that rules the world.'"

REPLAY ANSWERS

REPLAY—THE FRESH START

1. Standard English 2. F 3. F 4. F 5. F 6. F 7. F 8. T

REPLAY 1

Sentence No.	Noun	Verb
1	secretaries	type
2	Students	think
	careers	
3	manager	have
	accountant	
	reports	
	subject	
4	Fred	is
	president	
	company	
5	Ms. Adams	was
	manager	
	store	

Sentence No.	Noun	Verb
6	clerks	were typing
	answers	
	blanks	
7	salesperson	will have
	work	completed
	Tuesday	
8	Martha	should go
	conference	
	Kansas	
9	wife	had entered
	manager	
	lobby	
10	Mrs. Golden	will ask
	letter	

Sentence No.	Noun	Verb	Sentence No.	Noun	Verb
11	instructions	have been	17	Aldrich Company	received
	convenience	provided		check	
12	Ed	knows	18	auditor	checked
	lawyer			books	
	building			accuracy	
13	Reuben Singer	will present	19	Frankenstein	built
	actor			factory	
	concert			city	
14	Joe	should mail		moon	
	tickets		20	office	have
15	paper	is		secretaries	
	box			typewriters	
	window				
16	bookkeeper	will mail			
	invoice				

REPLAY 2

Part A

1. I me my myself mine
2. you your yours yourself yourselves
3. she her hers herself
4. he him his himself
5. we us our ours ourselves
6. it its itself
7. they them their theirs themselves

Part B

who	this	anyone	nobody
whoever	that	no one	everybody
whom	these	everyone	something
whomever	those	somebody	anything
whose	someone	anybody	everything
			nothing

REPLAY 3

Part A—noun underlined

2. an exclusive resort community
3. the beautiful Pacific
3. the crystal blue waters
3. the white sandy beaches
4. exquisite guest rooms
4. Spanish decor
4. gourmet dining
5. a heated pool
5. The view—serene, smogless, sunny

Part B

1. These, any
2. many
3. This
4. five, some
5. Several, this

Part C—Possible answers

cheerful	drab
old	new
clean	modern
pleasant	

REPLAY 4

Here are some possible answers.

1. today, immediately
2. well, poorly
3. carelessly, fast
4. carefully, accurately
5. so, exceptionally
6. especially, unusually
7. very, somewhat
8. too, very
9. too
10. always, never
11. home, there
12. more
13. usually, often
14. even, much
15. most, extremely
16. well-, beautifully, exquisitely
17. eventually, finally
18. so, very, extremely
19. not, never, rarely
20. probably, evidently, not, clearly

REPLAY 5

Part A—Possible answers

and	as	because	even though
but	if	when	while
or	although	until	so that
nor	before	since	whenever
for	unless	though	than
			after

Part B—Possible answers

around	toward
over	through
under	above
to	below
into	inside

Part C

1. nor
2. but
3. when
4. and
5. Although

Part D

1. in touch, with you, during the first part, of next week
2. on the behavior, of people, on their lifestyles
3. of media
4. for advertising, in several ways
5. at producing spot announcements, for their customers

REPLAY 6

Part A

1. verb
2. adjective
3. noun

Part B

1. noun
2. verb
3. adjective

Part C—Possible answers

1. I am typing a letter to Barbara Stern of Ardsley.
2. He earned an A in typing.
3. The typing paper is in the top drawer.

Part D—Possible answers

1. Please step this way.
2. Watch your step, please.

Part E

1. adjectives
2. Although I was born in the city, I now live in the country. (possible answer)

REPLAY 7

1.
(a) pronoun
(b) noun
(c) verb
(d) preposition
(e) adjective
(f) adverb
(g) verb

2.
(a) believe
(b) attempt

3.
(a) its
(b) it

4. noun

5.
(a) verb
(b) adjective
(c) noun
(d) conjunction
(e) pronoun
(f) pronoun
(g) noun

REPLAY 8

Part A

	SUBJECT	VERB
1.	Mr. Johnson	was
2.	You	introduced
3.	typewriters	are
4.	assignment	seems
5.	books	were
6.	Mr. Johnson	wants
7.	members	have
8.	We	walk
9.	secretary	types
10.	supervisor	found

Part B

1. V
2. S
3. S
4. S
5. V
6. S
7. V
8. S
9. V
10. V

Part C—These are possible answers; yours may be different.

1. Our experienced and competent bookkeeper usually eats hay every morning.
2. The former treasurer of this company runs for exercise every day.
3. The big black race horse in my stable is a very good typist.
4. A mink coat with a sable collar and a silk lining costs a great deal of money.
5. The little old lady in tennis shoes was caught stealing the sugar.

REPLAY 9

Part A

1.	F	6.	F
2.	C	7.	F
3.	F	8.	C
4.	C	9.	C
5.	C	10.	F

Part B—Possible answers

1. If you do a good job on this project, you will get a raise in salary after the first of the month.
2. Although most corporations use the services of an auditor to examine the books, some errors may never be found.
3. Synthetic fabrics are generally easy to launder even though some stains never come out.
4. Before she became the manager, important papers were difficult to locate.
5. Now that Ms. Ramirez has completed the training period, she will start to work for the new treasurer.

REPLAY 10

Part A

1.	F	6.	F
2.	F	7.	S
3.	F	8.	F
4.	S	9.	S
5.	S	10.	S

Part B—Your sentences will, of course, vary from these.

2. The personnel department, which is on the third floor, is closed today.
3. Mr. McCarthy, feeling that he was right, went to the manager.
6. The staff, having been on vacation last week, felt rested and refreshed.
8. They are the members of the team that won all the games last year.

REPLAY 11

1. once. You	6. inaccurate. He	11. C	16. C
2. C	7. C	12. times. They	17. C
3. symptoms. They	8. career. They	13. C	18. C
4. C	9. C	14. Machines. It	19. wife. Only
5. aged. It	10. C	15. C	20. C

REPLAY 12

Part A

1.	R	Washington; then
2.	R	use; however,
3.	CS	symptoms; therefore
4.	CS	homework; for
5.	CS	inaccurate; he
6.	C	
7.	CS	job; also

Part B

8.	R	quickly, for
9.	C	
10.	R	floor, but
11.	CS	once, and
12.	C	
13.	CS	bones, but
14.	CS	errors, but
15.	C	

Part C

16. also, in addition, otherwise
    for example, in fact, then
    hence, moreover, therefore
    however, nevertheless, thus
17. T
18. T
19. T
20. F
21. independents
22. run-on or a comma splice

REPLAY 13

Part A

1. CS		6. R		11. C	*Although*
2. C	*Because*	7. C	*because*	12. C	
3. R		8. C		13. C	*Since*
4. C		9. R		14. R	
5. C	*Even though*	10. CS		15. C	*Although*

Part B

16. I do not believe in shouting myself hoarse when (because) the team is playing its best.
17. Because (since) he places big orders with us, the manager is offering him a special discount.
18. Even though (Although) the plant was operating on a 24-hour basis, top management refused to adopt a three-shift schedule.
19. Even though (Although) the professor explained that studying English is fun, the new student wouldn't believe it.
20. Even though (Although) the method of shipment was not vital, we gave it careful consideration.

REPLAY 14

2nd line: representative, Bob Robinson,
3rd line: month. He . . . you. or month; he . . . you.
4th line: Mr.
5th line: HANDBOOK. This
6th line: now. We
7th line: carefully.
9th line: COMMUNICATION.
10th line: This
11th line: month.
13th line: available. If
14th line: know. Best
15th line: season.
16th line: sentences. They . . . read.

REPLAY 15

1. Answers will vary.
2. Answers will vary.
3. guide words
4. Answers will vary.
5. They show the alphabetic range of the page.
6. spelling, syllables, pronunciation, definitions, parts of speech
7. etymology, usage, illustrations, synonyms, derivatives, plurals, capitalization
8. unabridged dictionary
9. desk or college, pocket
10. desk or college
11. pocket
12. pronunciation key
13. Answers will vary.
14. front matter
15. AHD: Biographical Entries, Geographical Entries, Abbreviations, Four-Year Colleges and Universities, Two-Year Colleges and Universities, Picture Credits, Index.

    WNCD: Abbreviations, Foreign Words and Phrases, Biographical Names, Geographical Names, Colleges and Universities, Signs and Symbols, Handbook of Style, Index.

    Answers will vary depending on the dictionary you are using.

REPLAY 16

1. 3
2. 1st
3. 3rd
4. 2nd
5. AHD: Jules Leotard, (1830–1870); WNCD: Jules Leotard, 1870, Fr. aerial gymnast
6. LEE a tard (say it)
7. AHD: pet; WNCD: bet. Answer will vary with other dictionaries.
8. AHD: one that is used in recovering or removing a wreck, especially a truck with a hoist and towing apparatus used in towing disabled or wrecked vehicles.
   WNCD: same
9. c
10. woolly
11. AHD: catalog; WNCD: catalog or catalogue

12. a
13. 1st
14. ONTRA pra NUR (say it): one who organizes, manages, and assumes the risk of a business (exact wording will vary)
15. ə; WNCD: abut; AHD: about, edible, gallop, circus, item

REPLAY 17

1. talkative: loquacious, wordy, garrulous, voluble, effusive, verbose, glib.
   evident: manifest, patent, distinct, obvious, apparent, plain, clear.
   vacillate: See synonyms given under hesitate.
2. | AHD | WNCD | |
   |---|---|---|
   | sŭt′l-te | ˈsət-ᵊl-tē | noun |
   | rash′ə-năl′ | ˌrash-əˈnal | noun |
   | frăn′chīz′ | ˈfran-ˌchīz | noun, verb |

   Symbols will vary with dictionary used.
3. Janus
4. AHD: ME secretarie, confidant<Med. Lat. secretarius, confidential officer< Lat. secretus, secret. WNCD: ME secretarie, fr. ML, secretarius, confidential employee, secretary, fr. L secretum, secret fr. neut. of secretus.
5. 947,154
6. insert space; transpose
   AHD: let it stand; WNCD: let marked text stand as set
7. Frankfort
8. CXXXVII (AHD: find under Roman numeral; WNCD: find under Roman numeral; then see Number table. Look up table for CXXXVII.)
9. 1874
10. shilling (AHD: See Index under "currency"; WNCD: See Index under "Money.")
11. Organization of Petroleum Exporting Countries (See Appendix for Abbreviations.)
12. ac/cu/rate, com/pa/ny, cor/re/spon/dence, vac/ci/na/tion
13. False
14. dogie: chiefly West
    anyways: AHD: nonstandard; WNCD: archaic and chiefly dial.
    critter: AHD: regional; WNCD: none
15. AHD: characteristic of spoken language or writing that seeks the effect of speech; informal; conversational; WNCD: conversational

REPLAY 18

1. F	10. T	19. dilemma	28. deductible
2. T	11. separate	20. villain	29. fulfilled
3. F	12. weird	21. conscience	30. height
4. T	13. recommend	22. pursue	31. irrelevant
5. T	14. accommodate	23. congratulate	32. mathematics
6. F	15. privilege	24. pronunciation	33. occasion
7. T	16. bachelor	25. embarrassed	34. occurred
8. T	17. superintendent	26. accidentally	35. ridiculous
9. T	18. persistent	27. existence	

REPLAY 19

Part A

1. allies	6. copies
2. alleys	7. journeys
3. itineraries	8. authorities
4. proxies	9. accessories
5. facilities	10. surveys

Part B

1. sopranos
2. dynamos
3. embargoes
4. pianos
5. heroes
6. egos
7. potatoes
8. cargoes, cargos
9. portfolios
10. mementos, mementoes

Part C

1. thieves
2. handkerchiefs
3. knives
4. tariffs
5. calves
6. halves
7. safes (Note that *saves* is a verb, not a noun.)
8. wolves
9. plaintiffs
10. wives

Part D

Line 1: Globetrotters
Line 2: States
Line 3: fans, games
Line 4: audiences
Line 5: routines

REPLAY 20

Part A

1. formulas, formulae
2. addenda
3. alumni
4. appendixes, appendices
5. bases
6. bureaus, bureaux
7. censuses
8. AHD: criteria, criterions; WNCD: criteria
9. data, datums
10. indexes, indices
11. mediums, media
12. geese, gooses
13. AHD: oxen; WNCD: oxen, ox
14. parentheses
15. diagnoses

Part B

1. S
2. S
3. P
4. S
5. P
6. S
7. P
8. P
9. P
10. S

REPLAY 21

Part A

1. corps
2. politics
3. AHD: deer;
   WNCD: deer, deers
4. Marys
5. series
6. Fifes
7. aircraft
8. fish, fishes
9. DeSotos
10. statistics

Part B

1. P
2. S
3. S
4. P
5. S

Part C

1. have
2. is
3. are
4. were
5. are
6. are
7. is
8. was
9. has
10. were

REPLAY 22

Part A

	SINGULAR	PLURAL
1.	spoonful	AHD: spoonfuls WNCD: spoonfuls, spoonsful
2.	textbook	textbooks
3.	microprocessor	microprocessors
4.	editor in chief	editors in chief
5.	stockholder	stockholders
6.	AHD: passer-by passerby WNCD: passerby	passers-by passersby passersby
7.	businesswoman	businesswomen
8.	brother-in-law	brothers-in-law
9.	timetable	timetables
10.	bill of lading	bills of lading

Part B

1. general manager
2. C
3. manager, shipping department
4. C
5. Secretary of State
6. C
7. Office Manager
8. north, spring
9. city, state
10. English, Spanish
11. uncle, business, colleges
12. black, company
13. C
14. C
15. Nobel Prize

REPLAY 23

These are possible answers; yours may vary somewhat.

1. The women in my office go to lunch at 12.
2. He promised to send his secretary over with the contracts.
3. The director of the organization is a black woman.
4. Are you a feminist?
5. After dinner the men discussed the election while the women had their committee meeting.
6. We have several police officers guarding against intruders.
7. He is studying to be a nurse.
8. There's a lawyer in the office across the hall. (If the lawyer's sex is relevant to the information you are transmitting, it's acceptable to refer to her as a "woman lawyer.")
9. The author of the book you are now reading doesn't like to be called an "authoress."
10. The head of the committee is a bookkeeper.

REPLAY 24

1. timetable
2. sisters-in-law
3. alumni
4. plaintiff
5. criterion
6. criteria or criterions
7. allies
8. alley
9. cargo
10. bill of lading
11. portfolio
12. write-off
13. proceeds
14. itinerary
15. memento
16. chassis
17. chassis
18. proxy
19. hypothesis
20. embargo

REPLAY 25

1. me
2. me
3. he
4. us
5. I
6. We
7. me
8. him
9. we
10. she
11. me
12. us
13. They
14. him
15. He and I
16. me
17. her
18. We
19. me
20. me
21. me
22. he
23. him
24. me
25. they

REPLAY 26

Part A

1. C
2. themselves
3. he
4. C
5. me
6. he
7. I do (Add "do" because it "sounds right.")
8. C (Add "does" because it "sounds right.")
9. himself
10. I
11. I
12. ourselves
13. C
14. C
15. C
16. ourselves
17. themselves
18. himself
19. C
20. C

Part B

1. (does)
2. (he loves)
3. (does)
4. (I know, he knows)
5. (can)

REPLAY 27

1. who's or who is
2. C
3. C
4. C
5. ours
6. Yours
7. C
8. theirs
9. You're or You are
10. yours
11. They're or They are
12. one's
13. C
14. Who's or Who is
15. anyone's, theirs

REPLAY 28

1–7. jury, club, class, herd, flock, team, group, crowd, company, committee, staff, union, family, name of any organization (possible answers)

8. a
9. b
10. a
11. its
12. it
13. its
14. its
15. their
16. its
17. its
18. its
19. its
20. it

REPLAY 29

1. Whom (you would like *him* to accompany me)
2. whom (I met *him* at)
3. Whom (you do prefer *him*)
4. who (*he* needs it)
5. who (*he* [is] needed)
6. whom (I took *him* to be)
7. whomever (the convention chooses *him*)
8. who (*he* will do the)
9. Whoever (*He* is willing to do)
10. Whom (you would like *him* to go)
11. who (*he* must make)
12. whoever (*he* gets there)

13. who (*he* was chosen)
14. whoever (*he* needs it)
15. whom (you sent for *him*)
16. who (*he* would be best)
17. whom (we met *him*)
18. who (*he* helped me)
19. whom (I told you about *him*)
20. who (*he* has been active)
21. Who (*he* will get)
22. Whom (we should get *him*)
23. who (*he* should handle)
24. whoever (*he* can identify it)
25. whoever (*he* makes the most)

REPLAY 30

1. No one
2. anyone
3. Everybody
4. someone
5. Any one
6. no body
7. his
8. his
9. her
10. someone, he
11. somebody, his, he
12. Every one
13. Everyone, his or her
14. his
15. Everybody, his or her
16. Best: (c) Wrong: (b)
17. Best: (a) Wrong: (d)
18. Best: (c) Wrong: (b)
19. Best: (b) Wrong: (c)
20. Best: (c) Wrong: (b)

REPLAY 31

1. types
2. needs
3. arrived
4. are sailing
5. waxed
6. paints
7. consults
8. climbed
9. gained
10. want
11. wants
12. wanted
13. will want
14. offered
15. will select or should select
16. looks
17. looked
18. will consider or should consider
19. is considering
20. wanted
21. stay
22. are staying
23. watched
24. need
25. talked

REPLAY 32

1. broken
2. begun
3. chooses
4. chosen
5. does
6. C
7. eats
8. did
9. fly
10. gave
11. gone
12. risen
13. run
14. worn
15. speaks
16. saw
17. C
18. wear
19. seen
20. C
21. C
22. rang
23. C
24. C
25. give
26. is
27. is
28. are
29. are
30. is
31. are
32. flow
33. knows
34. are
35. is

REPLAY 33

1. biting, bit, bitten
2. blowing, blew, blown
3. .bursting, burst, burst
4. coming, came, come
5. costing, cost, cost
6. drawing, drew, drawn
7. falling, fell, fallen
8. forecasting, forecast or forecasted, forecast or forecasted
9. forgetting, forgot, forgotten
10. freezing, froze, frozen
11. grinding, ground, ground
12. hiding, hid, hidden
13. leading, led, led

14. paying, paid, paid
15. shaking, shook, shaken
16. sinking, sank, sunk or sunken
17. singing, sang, sung
18. throwing, threw, thrown
19. winning, won, won
20. writing, wrote, written
21. beaten
22. did
23. broken
24. hung
25. losing

(A few of these answers may vary, depending on the dictionary you're using.)

REPLAY 34

1. We were
2. They are or They're
3. are you
4. We have been or We've been
5. We are or We're
6. will be helpful or is helpful
7. We are or We're
8. C
9. C
10. You are or You're
11. Were you
12. was happy
13. is planning
14. C
15. were misplaced
16. were late
17. salespeople are
18. has been meeting
19. C
20. Were you
21. were
22. is, was, or will be
23. C
24. is, will be, or was
25. was

REPLAY 35

	SUBJECT	VERB
1.	I	enjoy
2.	salespeople	are doing
3.	you	do read
4.	He, I	did, could find
5.	We, he	voted, was elected
6.	we	won't issue
7.	personality	is
8.	Advancement, who	comes, do
9.	work	turns
10.	Playing	is
11.	sales	have risen
12.	students	study
13.	Suresh, who	is, presented
14.	typewriters	are
15.	Rachel Rothstein	will receive
16.	Everyone	is working
17.	A. B. Dick & Co.	manufactures
18.	assistant, who	will help, is
19.	you	would like
20.	turnover	is
21.	poverty, riches	are
22.	English Shoe Company	distributes
23.	We, you	are returning, hope, will take
24.	typewriters, which	will be delivered, have
25.	Chan & Lopez, Inc., name	has, has been placed

REPLAY 36

	SUBJECT	VERB	NUMBER
1.	groups	were invited	P
2.	She	wants	S
3.	family	took	P
4.	family	piled, left	S

	SUBJECT	VERB	NUMBER
5.	Abel & Lopez, Inc.	has	S
6.	Everyone	was waiting	S
7.	Everybody	seems	S
8.	report, letter	were	P
9.	report, letter	was	S
10.	principal, assistants	are	P
11.	story, story	sound	P
12.	story, story	sounds	S
13.	battery, radio, antenna	are	P
14.	everything	is	S
15.	Mr. Jones, Ms. Goldberg	will work	S
16.	person	rides	S
17.	class	took	P
18.	jury	arrived	S
19.	one	can do	S
20.	lawyer, layman	should be	P
21.	Half	is	S
22.	Half	are	P
23.	All	are	P
24.	All	is	S
25.	Most	are	P

REPLAY 37

	SUBJECT		VERB
1.	groups	1.	were invited
2.	She	2.	doesn't want
3.	people	3.	were
4.	Chan & Lopez, Inc.	4.	has
5.	figures	5.	are
6.	report, letter	6.	C
7.	report	7.	has been completed
8.	he, they	8.	C
9.	he, assistant	9.	wraps
10.	machines	10.	C
11.	Miss Adams	11.	arrives
12.	principal, assistant	12.	drive
13.	He	13.	does
14.	typewriter	14.	doesn't work
15.	cousin	15.	works
16.	accountant, manager	16.	C
17.	Mr. Perez, I	17.	give
18.	models	18.	C
19.	which (typewriters)	19.	have
20.	class	20.	C
21.	Roberta Simon	21.	deserves
22.	carton	22.	C
23.	man, woman	23.	wants
24.	All	24.	were mailed
25.	One-third	25.	have been sold

REPLAY 38

1. C
2. were instead of was
3. C
4. were instead of was
5. were instead of was
6. C
7. were instead of was
8. was instead of were
   was instead of were
9. were instead of was
10. were instead of was

REPLAY 39

1. *those kinds* of calculators or *that kind* of calculator
2. Those or These instead of Them
3. *these types* of errors or *that type* of error
4. *Those sorts* of books *are* extremely interesting. or *That sort* of book *is* extremely interesting.
5. *these types* of people or *that type* of person
6. He wants to buy this book.
7. that calculator
8. tell those people
9. *That is* the way or *That's* the way
10. *These kinds* of advertisements or *that kind* of advertisement doesn't

REPLAY 40

1. an	2. a	3. an	4. an
5. a	6. an	7. a	8. an
9. a	10. an	11. a	12. an
13. a	14. a	15. an	16. an
17. an	18. a	19. an	20. a
21. an	22. an	23. an	24. an

REPLAY 41

1. a, an
2. A, an, an
3. an, a, an
4. a, an
5. A, an
6. An, an
7. A, a, a
8. a, an
9. a, a, an
10. a, a, an

REPLAY 42

Part A

1. comparative: farther or further
   superlative: farthest or furthest
2. comparative: worse
   superlative: worst
3. comparative: littler or less
   superlative: littlest or least
4. comparative: more
   superlative: most
5. comparative: better
   superlative: best

Part B

1. bigger
2. farther
3. safest (omit "most")
4. worse
5. C
6. oldest
7. more accurate
8. heavier (omit "more")
9. more recent
10. wider
11. most beautiful
12. worse
13. less popular
14. brighter
15. friendliest

REPLAY 43

Part A

UNDERLINED ADJECTIVE	CIRCLED WORD	ANSWER
1. quick	walked—verb	quickly
2. bad	fumes—noun	C
3. clear, correct	should write—verb	clearly, correctly
4. beautiful	Mrs. Adams—noun	C
5. bad	sister—noun	C
6. careful	do—verb	carefully
7. legible	wrote—verb	legibly
8. delicious	pie—noun	C
9. deep	should think—verb	deeply
10. fair	will treat—verb	fairly
11. good	works—verb	well
12. sure	wish—verb	surely
13. good	know—verb	well
14. logical	speaks—verb	logically
15. calm	he—pronoun	C

Part B

16. again
17. back to you
18. together
19. s
20. s

REPLAY 44

1. logically
2. more quickly
3. more calm or calmer
4. worst
5. surely
6. better
7. C
8. more quickly
9. especially well, really well, very well, etc.
10. beautiful
11. worse
12. distinctly
13. more smoothly
14. C
15. more clearly
16. worse
17. best
18. gracefully
19. C
20. satisfactorily

REPLAY 45

1. I rarely, I hardly ever, or I seldom do
2. any or a instead of no
3. any instead of none
4. anymore instead of no more
5. C
6. doesn't know anything
7. won't go anywhere
8. would never or wouldn't ever
9. anymore
10. a or any instead of no
11. isn't ever or is never
12. hardly ever saw or rarely saw
13. could hardly
14. hardly any
15. Everybody likes

AY 46

CIRCLED NOUN	UNDERLINED NOUN	CIRCLED NOUN	UNDERLINED NOUN
1. C		12. C	
2. editors	stories	13. Californias	vineyards
3. C		nations	wine
4. brother-in-laws	manager	14. McCarthys	invention
5. C		15. Mens	College
6. attorneys	offices	16. C	
7. South Dakotas	resources	17. secretarys	book
8. Mens, womens	clothes	18. weeks	work
9. Smiths	orders	19. Districts	classrooms
10. industrys	directors	20. Ellens	visit
11. crews	strength		

REPLAY 47

Part A

SINGULAR	PLURAL
1. freshman's	freshmen's
2. week's	weeks'
3. witness's	witnesses'
4. James's	Jameses'
5. country's	countries'
6. secretary's	secretaries'
7. clerk's	clerks'
8. mouse's	mice's
9. wife's	wives'
10. father-in-law's	fathers-in-law's
11. congresswoman's	congresswomen's
12. manager's	managers'
13. fox's	foxes'
14. hour's	hours'
15. Wolf's	Wolfs'
16. wolf's	wolves'
17. principal's	principals'
18. boss's	bosses'
19. lady's	ladies'
20. child's	children's

Part B

1. boy's
2. members
3. communities
4. years'
5. person's
6. men's fashions, women's
7. company's assets
8. editors in chief's remarks
9. minute's
10. Perkins', days
11. ladies' coats
12. Goldsteins, days
13. San Diego's, years
14. Jones's
15. chairman's, details, workers'
16. Keats's
17. Knox's
18. Fritz's
19. Hendrix's
20. guests' names

REPLAY 48

1. couldn't, o'clock
2. i's
3. C
4. C
5. It's
6. Don't, R.N.s'
7. C
8. and's, if's, but's
9. '65
10. '29
11. C
12. '90s
13. l's, they'll, t's
14. Won't
15. A's
16. C
17. What's
18. You've
19. Aren't
20. Couldn't

REPLAY 49

1. gas, light, industries, towns,
2. C
3. desk, trash,
4. report, accuracy,
5. Chevrolets, Fords, Plymouths, Chryslers, Cadillacs,
6. 12, 6,
7. C
8. objectives, them,
9. first, second,
10. Machines, Motors, Telegraph,

REPLAY 50

1. C
2. pleasant,
3. came, wanted,
4. small,
5. C
6. Paris, Brussels,
7. courteous, helpful,
8. healthy, wealthy,
9. C
10. settled,

REPLAY 51

1. carefully,
2. today,
3. us,
4. C
5. division,
6. payment,
7. Rochester,
8. creative,
9. Miami, Dade,
10. C

REPLAY 52

1. now,
2. attractive, Street,
3. know,
4. report,
5. C
6. salesperson,
7. No,
8. Lee, washer, dryer, toaster,
9. C
10. accountant,
11. C
12. quickly,
13. time,
14. factory,
15. C

REPLAY 53

1. office, however,
2. costs, you,
3. employees, Brown,
4. office, 105,
5. Department, floor,
6. Simon, accountant,
7. C
8. C
9. work, therefore,
10. proposal,
11. today,
12. equipment, yet,
13. tomorrow,
14. C
15. will, believe,
16. George,
17. C
18. C
19. C
20. worth,
21. C
22. booklet, information,
23. language,
24. team, engineers,
25. C

REPLAY 54

1. 9, 1984,
2. C
3. Smith, Ph.D.,
4. Beach, York,
5. Friday, 4, 1984,
6. C
7. Rock, Arkansas,
8. Stores, Inc., Plains, York,
9. Ltd., Edinburgh, Scotland,
10. C

REPLAY 55

1. See samples (a), (b), (c), and (d) in Read Unit 55.
2. See Note 4 in Read Unit 55.
3. (a) no (b) no

REPLAY 56

1. p.m.,
2. quiet,
3. you, however,
4. 16, 1980,
5. No,
6. know," said,
7. very, very
8. morning,
9. run, run
10. excellent,

REPLAY 57

1. Use a comma after an introductory dependent that has a verb.
2. Use commas to separate parts of addresses.
3. Use a comma before *but* when two independents are joined.
4. Counting from the right, use a comma after each group of three numbers.
5. Use commas to separate items in series.
6. Use a comma in place of an important word that is omitted.
7. Use a comma between a date and a year and after the year.
8. Use a comma after an introductory expression that contains any verb form.
9. Do not use a comma around essential expressions.
10. Use a comma after the introductory expressions Yes and No.

REPLAY 58

1. conference.
2. day. hard. or!
3. express.
4. convention?
5. decisions. out.
6. convention.
7. you?
8. now!
9. letters.
10. Wonderful!
11. mail.
12. come?
13. come.
14. fire!
15. day!

REPLAY 59

1. stay;
2. C
3. year;
4. C
5. furniture;
6. ambition;
7. of;
8. Illinois;
9. C
10. understand;
11. C
12. low;
13. dividend;
14. 1902; 1913;
15. C
16. luck;
17. C
18. C
19. kills;
20. late;

REPLAY 60

1. following:
2. life:
3. C
4. C
5. cities:
6. shoplifting:
7. Valley:
8. about:
9. duty:
10. writes:
11. yesterday: secretary.
12. inefficient;
13. qualifications:
14. qualifications?
15. Office!

REPLAY 61

1. (I'll check the address)
2. thesaurus--a treasury of synonyms, antonyms, parallel words, and related words-- (or parentheses)
3. machines--they are the best money can buy-- (or parentheses)
4. charm--
5. C
6. attributes:
7. once--
8. skills:
9. (not $156)
10. live--a
11. ($100,000) or dashes
12. bookkeeping--
13. attractive--
14. (not now at least)
15. government--the executive, the legislative, and the judicial--

REPLAY 62

1. He shouted, "Your house is on fire!"
2. "Your house is on fire!" he shouted.
3. She whispered, "Are you sure you love me?"
4. "Are you sure you love me?" she whispered.
5. Do you know whether he said, "I love you"?
6. After he scolded her, he said, "What a beautiful day!"
7. Do you know what the phrase "negotiable instrument" means?
8. The following information is quoted from the letter: "The train leaves Chicago at two o'clock and arrives in Boston, Massachusetts, at midnight."
9. Was it Mr. Higgins who said, "Results are what count"?
10. It's important that they don't think you are a "pushy" sales representative.
11. Rosemary Wrenn of Long Beach said that she would like the job.
12. C
13. Alexander Smith, who lives in Toronto, will be writing the screenplay for us.
14. Please tell the chef how many tacos you need for the company party.
15. "This shipment of stationery," the manager clearly stated, "will arrive on time."
16. I have this to say regarding his "abject poverty": "It is fictitious."
17. Kristen said, "Are you all right?"
18. "M*A*S*H" was Ms. Johnson's favorite program.
19. "Bach Suite No. 2 in B Minor" is first on the program.
20. Ms. Crock whispered, "Did you know that the item marked 'fragile' was broken on arrival?"

REPLAY 63

Part A

1. I work in a ten-story building.
2. I need an up-to-date report.
3. My father is a hard-working man.
4. She made an off-the-record comment.
5. The case against the Dallas-based company was handled in Seattle.

Part B

1. re-creation
2. C
3. C
4. C
5. father-in-law
   commander in chief
6. self-made
7. self-sufficient
8. twenty-one.
9. C
10. anti-intellectual (An unabridged dictionary may be necessary to find this word.)

Part C

func/tion	be/lieve	hori/zontal	wouldn't/
thou/sands	punc/tu/ation	aligned/	im/pos/sible
in/ter/rupt	syl/lables	stopped/	guess/work

REPLAY 64

Answers will vary.

INDEX

1rst Spelling Test	92 1/2
Quiz 1	92
2nd Spelling	70
Quiz 2	68
3rd Spelling	92
Quiz	64
4th Spelling (Section Five)	92
Quiz	88 (4) credit #6
5th Spelling	100
Quiz	92
6th Spelling	100
Quiz	100
7th Spelling	100
Quiz	80
8th Spelling	100

Spelling 746.5 ÷ 8
93.3

Quiz 584 ÷ 7
83.4